The Sociology of Medical Science and Technology

***Sociology of Health and Illness* Monograph Series**

Edited by Jonathan Gabe
Department of Social Policy and Social Sciences
Royal Holloway
University of London

Current and forthcoming titles:
Medicine, Health and Risk
Edited by Jonathan Gabe

Health and the Sociology of Emotions
Edited by Veronica James and Jonathan Gabe

The Sociology of Medical Science and Technology
Edited by Mary Ann Elston

The Sociology of Health Inequalities
Edited by Mel Bartley, David Blane and George Davey Smith

The Sociology of Medical Science and Technology

Edited by Mary Ann Elston

Blackwell Publishers/Editorial Board

ISBN 0-631-20447-4

First published in 1997

Blackwell Publishers Ltd
108 Cowley Road, Oxford OX4 1JF, UK
and
350 Main Street
Malden, MA 02148, USA

British Library Cataloguing in Publication Data

A CIP catalogue record for this book is available from the British Library

Library of Congress Cataloging-in-Publication Data

applied for

Printed in Great Britain by Page Bros, Norwich, Norfolk.
This book is printed on acid-free paper.

Acknowledgements

I would like to thank all those involved in the various stages of producing this monograph. I am grateful to the individual contributors for generally responding swiftly and graciously to deadlines, comments and queries. The anonymous referees who read each of the chapters, usually two or three times, have helped the contributors and myself enormously. I am especially indebted to Jonathan Gabe, as Monograph Series Editor and colleague, as well as to Peter Conrad, visiting professor of sociology at Royal Holloway during 1996–7, and Mike Bury, for support and advice, Anthea Holme's copy-editing and Susan Gregory's administrative and proof-reading skills have been invaluable. The helpfulness and efficiency of the staff at Blackwell's have been much appreciated throughout.

Acknowledgments

[illegible]

Contents

Introduction: the Sociology of Medical Science and Technology 1
Mary Ann Elston

Science and Clinical Practice

1. Medical pedigrees and the visual production of family disease in Canadian and Japanese genetic counselling practice 29
Yoshio Nukaga and Alberto Cambrosio

2. Science versus care: physicians, nurses and the dilemma of clinical research 57
Mary-Rose Mueller

3. Bodies of knowledge: lay and biomedical understandings of musculoskeletal disorders 79
Helen Busby, Gareth Williams and Anne Rogers

Constructing Professional Tools

4. The rhetoric of prediction and chance in the research to clone a disease gene 101
Paul Atkinson, Claire Batchelor and Evelyn Parsons

5. Vital comparisons: the social construction of mortality measurement 127
Mel Bartley, George Davey Smith and David Blane

Assessing and Regulating Medical Technologies

6. The science and politics of medicines regulation 153
John Abraham

7. 'Strange bedfellows' in the laboratory of the NHS? An analysis of the new science of health technology assessment in the United Kingdom 183
Alex Faulkner

Notes on Contributors 209

Index 211

Introduction: the Sociology of Medical Science and Technology

Mary Ann Elston

In July 1997 plans were announced for a new biomedical research centre at University College, London (UCL), to investigate 'ways of treating and preventing heart attacks, strokes and cancer' (*Times Higher* 4 July 1997). The institute's director, previously director of research at a major UK-based international pharmaceutical company, described the new initiatives as 'a unique experiment in the organisation and practice of research into chronic adult diseases'.

> We are not just intending to study the problems, we also want to develop chemicals that can alleviate or prevent them.
>
> (*Times Higher*, 4 July 1997).

This new institute may be unique in some respects but its establishment also illustrates some general themes germane to the sociological study of medical science and technology and to this monograph. For example, the research and financial collaboration between university and corporate sector exemplifies the increasingly hybrid character of scientific research, particularly but not only in the biomedical field. A sharp distinction between pure and applied science and their associated scientific institutions looks decreasingly tenable. Indeed, the explicitly multi-disciplinary, heterogeneous, problem-solving approach of the new institute exemplifies the new mode of knowledge production in the context of application (Mode 2) identified by Gibbons *et al.* (1994). But the new institute is clearly premised on the assumption that the application of laboratory-based biomedical science and chemical technologies is a key route for improving health, whether this is via treatment or preventive interventions. It is an institutional expression of a conceptual approach to health care that medical sociologists have often tagged the 'biomedical model' (*e.g.* Nettleton 1995), or the 'cosmology of laboratory medicine' (Jewson 1976).

The development of medicines is the primary aim of the new institute. Medicines are, according to Davis, 'the personal technology of our times' (1997: 1). They are, at least in the developed world, in virtually universal use, carrying powerful cultural associations, incorporating the achievements of modern science and underpinning the work of medicine and pharmacy and the foundation of one of the most successful manufacturing industries. Questions of efficacy and safety of these powerful technologies raise 'a wide range of regulatory and policy issues that go to the heart of the modern welfare state' and 'to the heart of globalisation processes' as the regulation of medicines transcends national frontiers (Davis 1997: 1). Such regulatory issues bring scientists, industrialists, politicians and administrators (and sometimes consumers) together in complex networks of state and transnational advisory and executive bodies.

The site for the new institute has rich historical associations. UCL played an important part in the slow rise of laboratory medicine in Britain from the mid-nineteenth century. On the opposite side of the road from the original college buildings stands the Victorian building built for its associated teaching hospital, University College Hospital (UCH). This building will house the new institute. The history of the complex, changing relationships between UCL and UCH is the history of the complex, changing relationships between biomedical science and hospital medicine in microcosm (*e.g.* Bynum 1994, Lawrence 1994). And this history is expressed in their spatial configuration. On the surface, Gower Street, a major thoroughfare by the late nineteenth century, has long divided the basic scientists from the clinicians, marking the separation often found in their professional lives and the boundary between two kinds of knowledge (Berg 1995). But, generally unknown to the passing public, the old hospital building (and presumably the new institute), the clinical medical school and the biomedical sciences buildings on the other side of Gower Street are connected by subterranean tunnels. Through the tunnels have passed generations of medical students, their teachers, hospital and research staff and bodies, dead and living, for teaching and research purposes. Whether or not the new UCH being built a short distance away will be linked to the College by large subterranean tunnels, it is a safe bet that the electronic cable connections will be extensive: data, images, virtually real bodies, will pass invisibly beneath the streets.

The subterranean connections provide us with a metaphor. Underneath modern health care practice is an extensive network of links to laboratories, scientific disciplines and industry. As Atkinson has pointed out, 'a great deal of medical work goes on away from the patient and the consultation' (1995: 1). Much of this work is normally either invisible to users of health care or packaged in the more or less taken for granted technologies that are an integral part of the routines of clinical practice (Berg 1992): for example, taking tissue samples for tests, the use of monitoring and investigative machines, and the prescribing of medicines.

Whether or not all this is the best possible use of scarce resources for improving the population's health is not the primary concern here, important issue though this is. Nor is there space for an historical account of how this situation came to be.[1] The point being made here is a simpler one: to draw attention to the vast array of institutions and practices in modern society that articulate the application of scientific method to the study and management of human health. A very significant proportion of those institutions and practices deemed 'scientific' in modern societies are concerned with health and health care. Changes in the way biomedical research work is organised and technologies developed, as with the new UCL institute, may have significant implications for individuals' careers, for the shaping of research agendas (*cf.* Webster 1994) and, of course, for health and health care.

However, the subterranean metaphor was not intended to imply that the relationship between 'science' and 'health care was necessarily one of science underpinning health care; that is, a relationship in which 'pure' knowledge is created by scientists in laboratories, developed into technologies and then applied by clinical or public health practitioners to lay people's health problems. To the patient, the dependency may often look more like the converse: relatively lowly laboratory technicians use routine technologies to serve powerful clinicians. Information, knowledge and technologies move in both directions between practice and the laboratory. The boundaries between different kinds of 'laboratories' may be permeable, the distinction between science and health care practice blurred. As Berg (1995) has shown, there have been diverse conceptualisations of the relationship between science and medical practice and of medical practice as (ideally) a science and, hence, of its alleged shortcomings. As several contributions to this volume will show, to write of 'linking' science and medicine as if they took place in different settings may sometimes be misleading. The 'bedside', the clinical setting, may be a source of data used simultaneously for research and treatment, as when pedigrees are constructed in genetics clinics (Yukaga and Cambrosio, this volume) or primarily a site for research, as in clinical trial units (Mueller this volume). There are other forms of medical science besides the experimental biomedical disciplines, including, arguably, the social sciences applied to medicine and the growing bodies of practitioner-developed knowledge, for example in nursing (Witz 1994). Clinical practice itself is increasingly both the phenomenon to be studied and a 'laboratory' for the emerging 'science' of health services research (HSR) (*cf.* Maynard and Chalmers 1997, Faulkner, this volume). The recent expansion of Mode 2 knowledge production (Gibbons *et al.* 1994) throughout modern health care is striking (but not uncontested).

Thus, the complex and multi-faceted relationships between medicine and science in particular contexts and the discourses that construct these relationships might be regarded as topics for sociological enquiry. There are also two flourishing subfields of sociology which might seem well-equipped to take up these topics: the sociologies of medicine and of science.[2] However, until very recently, the study of medical science and technology has not figured very prominently in either subfield.

On the one hand, (and with some notable exceptions) much of the relatively invisible, backroom world of the laboratory and its associated regions has, historically, received little attention in mainstream medical sociology. The arrays of tests and technologies that constitute much of contemporary medical practice have seldom been problematised, except when represented as extensions of a Foucauldian clinical gaze. The extent to which the clinical gaze might itself have been supplanted by the laboratory has rarely been explicitly considered (*cf.* Atkinson 1995). New technologies may carry profound implications for the experience of illness and the organisation of care, as when new drugs reconstitute acute crises into chronic illness in the community (Bury 1997). But how these technologies are developed and introduced into practice has often been taken to be a technologically determined given. The 'biomedical' model has been much criticised by medical sociologists (see *e.g.* Nettleton 1995). But the workings of the model or biomedical scientists' response to such criticisms has been less often examined in detail (Elston 1994). In their recent call for bridge-building between the sociologies of science and medicine, Casper and Berg describe the erstwhile stance of medical sociologists in a sharp tone. 'The investigator stood with his or her back to the heart of medicine and studied the "social phenomena" surrounding it' (1995: 397).

In contrast, sociologists of science and technology have produced many detailed studies of biomedical science in the laboratory (*e.g.* Latour and Woolgar 1986, Lynch 1985) and of the development of medical technologies (*e.g.* Yoxen 1987). Studies have followed biomedical science and scientists out of the research laboratory into policy-making and public controversy (Petersen and Markle 1981) and regulatory systems and advisory committees (Jasanoff 1990). But, until relatively recently, movements between the laboratory and the clinic and health care settings as a site for science received little attention. In 1988, Richards noted that 'while few, if any, areas of science have escaped the attentions of the revisionist sociologists of knowledge, medicine had remained curiously immune from their scrutiny' (1988: 654).[3]

These revisionists (see below) have adhered to Latour's maxim to study '*science in action* and not ready-made science or technology' (1987: 258). They have focused on 'core sets' of elite scientists (Collins 1985). These orientations have been enormously fruitful for studying controversy in expert, 'journal' science. But they have simultaneously rendered the world of ready-made 'textbook science' (Fleck 1935/1979) and routinely applied 'black boxed' technologies marginal to sociology of science.[4] Yet this is the science and technology with which most doctors, nurses, radiographers, dieticians and so on (and the public as patients) are mainly engaged. Whether and how 'experimental' medical technologies become more or less stabilised (or are contested) in practice settings or how not-very-expert (by the standards of research scientists) health care professionals mediate medical science and technology to the public or articulate medical research efforts are, arguably, pertinent issues.

Recently, however, there have been signs of growing interest in such issues and some rapprochement between the two subfields. For example, since Bartley's (1990) programmatic call, an increasing number of contributions explicitly applying sociology of science approaches to health care have been published in medical sociology journals such as *Sociology of Health and Illness* (*e.g.* Arksey 1994, Berg 1992, Mol and Elsman 1996, Prout 1996). Similarly, medical technologies and practice have recently figured prominently in key sociology of science journals (*e.g.* Berg 1995, Clarke and Montini 1993, Cussins 1996, Epstein 1995, Hartland 1996, Hirschauer 1991, Singleton and Michael 1993, Timmermans 1996) including in a special issue of *Science, Technology and Human Values* on medical practice (see Casper and Berg 1995). Edited collections and research monographs have recently been published on medical science and practice in general (*e.g.* Berg and Mol 1997) or on specific aspects: including the pharmaceutical and equipment industry (Abraham 1995, Blume 1991, Davis 1996, 1997), cancer research (Fujimura 1996, Richards 1991), haematology (Atkinson 1995), medical decision-making techniques (Berg 1997b), health economics (Ashmore *et al.* 1989), the 'new genetics' (*e.g.* Nelkin and Lindee 1995) and science in the politics of human reproduction (Mulkay 1997). A similar convergence of interests is detectable in the anthropologies (*e.g.* Good 1994, 1995, Traweek 1993) and histories of science and medicine (*e.g.* Löwy 1993, Pickstone 1992, 1993).

Emerging Contexts and Content for the Sociology of Medical Science and Technology

Several developments in and pertaining to medical science and technology over the last two decades have provided a context for and much of the content of this apparent rapprochement. They make this monograph particularly timely and are reflected in its content. One such development is the apparent resurgence, at the end of the twentieth century, of the threat of life-threatening epidemic infectious disease on a global scale. Since the early 1980s, a huge, heterogeneous network of scientific activity associated with the HIV/AIDS epidemic has developed, including a now considerable sociological dimension (see *e.g.* Bloor 1995). The latter has included some studies of the scientific activities which are highly pertinent to both sociology of medicine and of science (*e.g.* Epstein 1995, Fujimura and Chou 1994, Hart *et al.* 1992, Martin 1997). Given increasing claims in popular and professional media of a more general resurgence of epidemic disease (Bury 1997, Irwin and Wynne 1996b), epidemics and the scientific response to them might well be a focus for sociological research for the foreseeable future.

Another development is the rapid growth of molecular genetics and its application to human disease and in health-related biotechnologies. This 'new genetics' has been widely seen as having profound implications for society and for individuals (*e.g.* Lippman 1991, Marteau and Richards 1996, Richards 1993). This has been the starting point for much of the growing body of sociological research (*e.g.* Atkinson and Parsons 1992, Conrad 1997, Draper 1991, Lambert and Rose 1996, Nelkin and Lindee 1995).

There are also studies of the shaping of the new genetics knowledge and technologies (*e.g.* Balmer 1996, Fujimura 1996, Hilgartner 1995). Human disease genetics is, currently, a field in which laboratory science and clinical practice are particularly close. Because of the controversy that surrounds the field and because few applications have become routinised, sociologists of medicine cannot treat the technologies as given and sociologists of science cannot remain in the laboratory. It is no coincidence that this monograph contains two contributions on medical genetics.

The restructuring of biomedical research alluded to in the opening of this chapter has coincided with major reforms of health care systems in many parts of the world. Common elements include the strengthening of managerial authority over the organisation of clinical work and extending the evaluation of service effectiveness and efficiency (Ham, *et al.* 1990). Service needs have tended to be privileged over academic clinical research (Hafferty and Light 1995). But the clinical trials required in pharmaceutical research and development have to be accommodated within these reforms (Mueller, Abraham, this volume). And any constriction and restructuring of biomedical clinical research is being accompanied by the proliferation of HSR, of new, often social science-based knowledge and technologies, within the clinic. As with the HIV/AIDS epidemic and the 'new genetics', sociologists have begun to contribute to HSR and to subject some aspects of it to more detached scrutiny (*e.g.* Ashmore *et al.* 1989, Berg 1997a, Faulkner this volume).

The Sociologies of Medicine and Science

The preceding section has depicted two subfields of sociology with a gap between them that may now be closing. Of course, the boundaries between the two subfields are not rigid, any more than are the boundaries between the sociologies and anthropologies or histories of medicine or science (*cf.* Jasanoff *et al.* 1995). There are some longstanding common concerns and individuals who move between the two subfields. Nevertheless, there are identifiable professional communities and publishing outlets for each field and distinctive concerns and conceptual and methodological emphases which are still detectable within the recent work on medical science and technology just outlined. Some of these points of contact and departures are sketched here.[5]

Theoretical and methodological traditions that span the whole discipline of sociology provide one form of linkage. For example, from within the Marxist political economy tradition, there have been many critical accounts of science, technology and medicine as shaped by the social and material requirements of capitalism (e.g. Doyal with Pennell 1979, Levidow and Young 1982). Interactionism and ethnography have been central to both fields, with the inspirational influence of the late Anselm Strauss being widely acknowledged in both 'camps' (e.g. Atkinson 1995, 1997, Fujimura 1996, Star 1995). Before the two subfields were institutionalised, there were 'founding fathers' (the term is used advisedly) to be claimed by posterity in both fields. The American functionalist, Robert Merton, is one such. Merton's work is widely cited as establishing the study of science as a social institution, with its own distinctive prescribed norms (universalism, communality, organised scepticism and disinterestedness), as a distinctive field within sociology (e.g., Webster 1991, Storer 1971). Merton figures less prominently in general medical sociology textbooks today although his influence on another of that field's functionalist founding fathers, Talcott Parsons, has been recognised (Gerhardt 1989:357). But the 1950s study led by Merton (Merton *et al.* 1957) of an experimental programme in medical education still serves as a point of departure for most studies of the socialisation of health care professionals (e.g. Melia 1987).

'Point of departure' is apposite. From the late 1960s, much recent work in the sociology of science has located itself as 'post-Mertonian', the epithet intended to convey more than the passage of time (Bartley 1990, Webster 1991). A similar distancing from the functionalist approach to professions is evident in medical sociology. 'Post-Mertonian' sociology of science and medicine claims to provide a more critical analysis of medicine and science, rejecting any notion that these bodies of knowledge and those who develop and use them are 'outside society'. Before discussing this work, it is worth noting that Merton's own position was not as anodyne or as unmindful of the cognitive structure of science as some critical representations suggest (Macdonald 1995, Storer 1971). Moreover, the very significant contributions to sociological studies of medical science and technology made by close associates and students of Merton should not be overlooked. These include Renée Fox's many studies of medical and scientific uncertainty and clinical experimentation (*e.g.* 1959/1974, 1980, Fox and Swazey 1974) and Bernard Barber's work on therapeutic drugs and medical experimentation with human subjects (*e.g.* 1967, 1990).

Health and Scientific Professions and their Work

In the more sceptical perspective that emerged from the late 1960s, functionalism's acceptance of the medical and scientific professional community's prescriptive norms as starting points for the analysis of professional action was displaced by interactionist studies. These showed day-to-day scientific or medical work or education as being no different from, no more intrinsically altruistic or universalistic than, any other sphere of work or education (see Fujimura 1996). An emphasis on the function of the professions for society was displaced by more critical analyses of professions as agents of social control and of the political processes through which professional autonomy and power had been achieved (Macdonald 1995). Such critical studies of medicine and, latterly, other health care professions and their power in and out of the workplace have played a prominent part in medical sociology since the 1970s (see for *e.g.* Gabe *et al.* 1994).

Such explicit concerns with professions or occupations have been much less prominent in post-Mertonian sociology of science (Richards 1988). But there has recently been growing interest in Abbott's (1988) work on jurisdictional competition between professionals (*e.g.* Fujimura 1996). And many of the sociology of scientific knowledge (SSK) studies described below are, implicitly, studies of divisions of labour and inter- or intra-occupational interactions. Sociologists studying science in action have often adhered to Latour's (1987) instruction to follow scientists 'up' out of the laboratory to examine how (overwhelmingly male) scientific entrepreneurs and politicians garner the resources that make laboratory science possible (e.g. MacKenzie 1993). Rather less frequently, they have looked 'down' at those who do much of the mundane, unsung articulation work of science: the 'machine work', 'safety work' and, particularly where human or animal research subjects are concerned, the 'sentimental work' that make research possible. These include technicians and, in much medical research, nurses (*cf.* Rose 1994, Shapin 1989, Star 1995, Strauss *et al.* 1985, Mueller this volume).

The Construction of Medical Knowledge and Technology

The other, related point of departure from Merton has been the critical reappraisal of his omission of the content of science or medicine from sociological scrutiny (Bartley 1990, Webster 1991). In the context of post-Kuhnian philosophy of science (Kuhn 1970), sociologists and others turned their attention to the processes through which scientists make judgements about the value of scientific representations of nature. Fujimura's concise summary can serve for heuristic purposes here:

> These studies questioned [or often appeared to question – MAE] the assumption that judgments are grounded in some absolute truth or reality and compared scientific representations against one another by examining the methods and negotiations among heterogeneous elements used in the production of representations. They argued that science analysts should be agnostic and symmetric in their treatment of the cognitive elements of scientific representations (1996: 237).

Thus, advocates of the 'strong programme' for the sociology of scientific knowledge (SSK) (as opposed to the Mertonian 'weak' programme of sociology of science) focused particularly on the shaping of science by social interests (Bloor 1976/1981). Other SSK approaches adopted since the 1970s have included discourse and ethnomethodological analyses of scientific texts and talk (*e.g.* Gilbert and Mulkay 1984, Lynch 1985) and ethnographies of laboratory practice (*e.g.* Latour and Woolgar 1986). One feature that all these divergent strands within SSK share is that they make their general theoretical claims through empirical example, through painstaking description of scientists' practices and their representations of science. Labelling all these as 'social constructionism' conceals some very divergent views and controversies within SSK. Moreover, recent work in SSK has tended to drop 'social' in favour of 'constructionism' or 'constructivism' alone. To emphasise the 'social' appears to privilege appeals to social causes when appeals to nature have been deemed inadmissible. Moreover, the presupposition that the social can be distinguished *a priori* from the material or technical is rejected by proponents of the translation approach or actor-network theory (ANT) as developed by *inter alia*, Callon (1986), Latour (1987, 1988) and Law (1992) and drawn on in Yukaga and Cambrosio's and Bartley, Davey Smith and Blane's chapters in this monograph. ANT 'focuses on the strategies scientists use for building networks to make findings into facts' (Fujimura 1996:238). Scientists (or any actors, for it is a general theory about social life) are regarded as being multi-faceted entrepreneurs 'who engage in activities that might otherwise be deigned political, social or economic as well as those practices traditionally assigned the label "scientific" ' (Michael 1996: 53). The sociological task is to follow the actors wherever they go in constructing networks. Knowledge (or machines or social institutions and organisations) are seen as the product of this heterogeneous engineering, these networks of heterogeneous materials, human and non-human. (One of the more controversial aspects of ANT is its ascription of 'actor' status (although not intentionality) to non-humans such as equipment (see for *e.g.* Michael 1996).

Since the early 1980s, SSK has devoted increasing attention to technology as well as to science, with many practitioners refusing to make a distinction and referring only to 'technoscience' (see Bijker *et al.* 1987). As against longstanding sociological concerns with the social impact and implications of technology, they have generally rejected that formulation as implying asocial technological determinism; that is, the view that technological development follows an autonomous, technically shaped path which then shapes society. Studies of the social shaping of technologies (MacKenzie and Wajcman 1985) and social constructionist or, latterly, constructivist approaches (Bijker and Law 1992, Bijker 1993) have produced many detailed case studies of specific technologies, from nuclear weapons (MacKenzie 1993) to the bicycle (Bijker 1993). ANT's reconceptualisation of the technology-society relationship as a mutually constitutive one has been used in the close examination of technologies in daily use, such as metered dose inhalers for asthma (Prout 1996).

The 'social construction' of medical knowledge has also been a growing preoccupation of medical sociology over the last two decades, particularly since the publication of Wright and Treacher's (1982) *The Problem of Medical Knowledge* and exchanges between Bury (1986) and Nicolson and McLaughlin (1987, 1988). Work in SSK has been particularly influenced by philosophy of science and analytic philosophy, symbolic interactionism and ethnography. But the intellectual precursors of social constructionism in medical sociology have been even more diverse (Bury 1986, 1997). As well as post-Kuhnian philosophy of science, radical psychiatry, Marxism, critical theory and, in particular, the writing of Foucalt (*e.g.* 1976) have all been drawn on. More recently, the emerging sociologies of the body (Williams 1997) and of risk (Gabe 1995), have been influential in medical sociology's discussions of representations of medical knowledge. Thus 'social constructionism' in medical sociology is also a far from unitary phenomenon. But the prefix 'social' is still widely current in medical sociology's discussions of constructionism. As Casper and Berg (1995: 397) argue, much of its focus on medical knowledge is seen as a means to an end, 'as enhancing medical sociology's potential to trace relations of power'. Thus epidemiology is analysed in the Foucauldian perspective as a means of extending the clinical gaze into the community (Armstrong 1983). In contrast, most constructionism in SSK is, overtly at least, less concerned with the functions of knowledge for social control.

Medical sociologists' studies of medical technologies have tended to focus on the social implications of technology (*e.g.* Willis 1994) and only to a lesser extent, on its shaping by social interests or construction (*cf.* Prout 1996) with some studies of the diffusion of medical technologies into practice (McKinlay 1981, Press and Browner 1997). Some have drawn on the emerging sociology of the body, particularly post-modernist and post-structuralist ideas about 'cyborg culture' (Haraway 1991, Shilling 1993).[6] For example, Williams (1997) has argued that modern medical technologies, such as genetic engineering, bionics and cosmetic surgery, have rendered the human body as more plastic and more uncertain than in the past. The body becomes an unfinished project rather than a fixed entity. Here again, the emphasis is often less on the technologies themselves than on a romantic vision of their putative effects (*cf.* Schroeder 1996).

That the broad genre of sociological approaches just described, 'social constructionism' or 'constructivism', SSK, ANT or post-structuralism and so on, are contentious hardly needs stating. The debates about realism versus anti-realism, ontological and methodological relativism and the problem of reflexivity among those who adhere to one or other strand within this broad tradition have been too extensive to summarise here (*e.g.* Bijker 1993, Bloor 1991, Woolgar 1988) let alone consider the extensive criticisms from scientists (*e.g.* Wolpert 1992) and philosophers (*e.g.* Searle 1995). Some constructionists (but not all) would deny that their work implies a rejection of epistemic realism, the notion that there is a world that exists independently of human belief and language (a notion that Searle (1995) argues is a precondition of most intelligible discourse). But there remain major questions about the status of the knowledge claims of those who espouse 'only' methodological relativism (see Hollis and Lukes 1982). Readers who wish to pursue the sociological critiques of constructionism in sociology of science and medical sociology might turn to the work of Abraham (1995 and this volume) or Bury (1997). Abraham, in particular, has argued for the continued viability of a 'weak programme' in the sociology of scientific knowledge, a programme which he argues can escape the internalism and the neglect of questions of power and interests that, in his view, characterises much work in SSK. He suggests that the important empirical and methodological contributions of SSK have been achieved '*despite its declared relativism rather than because of it*. . . . The abandonment of relativism, therefore, does not entail the demise of sociology of scientific knowledge (1995: 27, emphasis original). Similarly, Schroeder, in his study of virtual reality technologies, has recently argued against non-realist constructionism but for a realist sociological analysis of technology that 'recognises that there are different types of interplay between technological and social forces on different levels, which means that in order to understand the social implications of new technologies, we must combine these levels' (1996: 9).

Lay and Expert Understanding of Science and Medicine

One of the attractions of a sociology of scientific or medical knowledge was that it promised a more autonomous role for the sociologist: a move not just from sociology *for* medicine/science to sociology *of* medicine/science from the margins but towards a sociology of the core of medicine/science (Bartley 1990: *cf.* Casper and Berg 1995). One might say that SSK has, itself, been a form of boundary work 'which occurs as people contend for, legitimate or challenge the cognitive authority of science – and the credibility, prestige, power and material resources that attend such a privileged position' (Gieryn 1995: 405, *cf.* Strong 1979). A similar trajectory (and the influence of constructionism) is detectable in the development of another area of cognate interest for the two subfields: the study of the lay public's ideas and understanding of medicine/science, of health and the workings of the body. Reviewing recent literature here, the overlap of concerns between the two subfields is striking as is, for the most part, the lack of cross-citation.

One starting point for sociological investigation of the public's ideas has been what is often labelled in sociology of science as the 'public deficit model'. This focuses on the gap between lay and authoritative scientific or professional understanding, as established, for example, in surveys testing the population's grasp of accredited scientific knowledge (see Irwin and Wynne 1996b, Michael 1996). As conceptualised in much health policy and the growing movement to promote the Public Understanding of Science (see Michael 1996), this deficiency on the part of the public has serious practical implications. Ignorance or irrational beliefs are alleged to lead to inappropriate behaviour on the part of the public: failure to comply with medical advice, failure to consult doctors appropriately, failure to abstain from risky behaviour and failure to become interested in scientific careers (see Bury 1997, Good 1994, Irwin and Wynne 1996b). It also, allegedly, leads to political problems: to ill-informed and irrational protests against legitimate scientific activities (see Elston 1994) and to a more general inability of the public to play their part as informed citizens in democratic debate (see Irwin and Wynne 1996b, Michael 1996). The remedy generally proposed is increasing the public's scientific or medical literacy: implicitly assuming that to know science/medicine is to love science/medicine.

That the general public falls short when measured by the standards of expert science/medicine is perhaps not surprising. Nor is this contested by sociological critics of the public deficit approach, such as Wynne in the sociology of science (*e.g.* 1991, 1996) or Gareth Williams in the sociology of medicine (*e.g.* Williams and Popay 1994, Busby *et al.* this volume). What these critics contest is the presumption that the universalist knowledge of expert science is necessarily the appropriate standard and that more exposure to it will necessarily convert a sceptical public. They argue instead for attention to the contextualised public understandings of science and medicine. The categorical distinction between an undifferentiated lay understanding and an equally undifferentiated expert one is replaced by a more relational view (Irwin and Wynne 1996b, Michael 1996). The local knowledge of workers, the experiential knowledge of sufferers from chronic disease, the variable distribution of expert knowledge among health professions and the challenges posed by lay activities to expert science may then be recognised (Wynne 1996, Grinyer 1995, Bury 1997, Collins, *et al.* 1998 f.c., Epstein 1995, Indyk and Rier 1993). The complex mixture of lay trust and lay scepticism towards modern medical science can then emerge (*e.g.* Williams and Calnan 1996, Lambert and Rose 1996).

Thus the relationships between expert and lay understandings of medicine and science, the construction of medical knowledge and technology and its social implications and the division of labour and organisation of work in medical science are all topics where the sociologies of medicine and science might fruitfully come together. And there is one area where all these overlapping interests and extensive cross-fertilisation between the sociologies of medicine and science (and other fields) have been evident for some time – the study of gender and medical science and technology.

Gender and Medical Science and Technology

Notwithstanding the continuing gender-blindness of much mainstream work in both sociology of medicine and sociology of science (*cf.* Delamont 1987), sociological studies focusing on gender and medical science have proliferated since the 1970s. The study of gender and medical science has been a lively 'trading zone' (Keller 1995: 80) not just between the sociologies of medicine and science but also between philosophy and history of science, literary theory, postmodernism, anthropology and feminist theory, with much input from reflective biologists and biomedical scientists (*e.g.* Hubbard 1990). Most of the work in this trading zone is engaged in a political or cultural critique of medical science as it currently is, sometimes profferring an alternative vision (see Rose 1994). Initially, much of it focused on women rather than gender although this has changed in recent years, especially in work informed by post-structuralist critiques (see Haraway 1991, Annandale and Clark 1996).

Since the 1970s, there have been three main themes in work on gender and science in general: studies of women in scientific work; science's construction of sexual differences or women's inferiority and the use of gender in science (*cf.* Keller 1995). Although there is an immense body of work falling under the first theme, its focus has been mainly on women's absence from expert science or discrimination against the few who did enter. This problematic has perhaps obscured the fact that, in most countries, the presence of women in biology, biomedical science and medicine has long been high relative to their presence in elite occupations in general and in other high status academic science disciplines (Elston 1993, Rose 1994, Schiebinger 1989). Women are also omni-present but invisible in scientific settings as cleaners and secretaries and so on. And, as indicated earlier, sociologists of science have been slow to recognise that nurses and paramedical professionals study and sometimes do science. In the last twenty years, sociologists of medicine have extensively examined the gendered division of labour in health care and the ways in which gender is a resource in organizational culture or tacitly incorporated into the construction of occupations (*cf.* Atkinson and Delamount 1990, Davies 1995, Riska and Wegar 1993, Witz 1994). But this kind of approach is much less apparent in the sociology of science. Thus the study of gender and work in medical science is relatively underdeveloped (Delamont 1987, Rose 1994).

In contrast, work addressing the other two themes in the gender and science literature is preoccupied with medicine and biomedical science. These fields have been conceptualised as patriarchal ideologies which legitimate women's inferiority in, for example, analyses of the representation of women (and men) in medical texts (Martin 1989). Medical technologies have been assessed in terms of their implications for control or liberation of women. And sociologists have examined how women, particularly through the women's health movement, have contested (or sometimes embraced) their development and use (*e.g.* Clarke and Montini 1993, Montini and Ruzek 1993). Literature in the third category draws on the constructionist, Foucauldian and post-modernist influenced perspectives on scientific and medical knowledge discussed earlier. It focuses on representations of biomedical science and medicine, on how these are saturated with gendered images and sexual metaphors and structured by gendered practices and how these in turn construct what it is to be female (or male) (*e.g.* Annandale and Clarke 1996, Haraway 1989, 1991, Jordanova 1989, Keller 1995, Martin 1994, Oudshoorn 1994, Pfeffer 1993, Rose 1994).

Perhaps the core activity in this trading zone of gender and medical science and technology is the study of human reproduction: that is, the science, technology and practice related to childbirth, contraception and abortion and, more recently, to conception and pre-natal screening and interventions. In particular, research on the so-called new reproductive technologies (NRTs) has produced fruitful exchanges between sociologists of science and medicine as well as with anthropologists, cultural theorists and sociologists of the family (*e.g.* Gray 1995, McNeil *et al.* 1990, Stacey 1992, Stanworth 1987). As with the 'new genetics', because of the moral, political and scientific controversy that surrounds NRTs (see for *e.g.* Clarke and Montini 1993, Mulkay 1997), it is hard for medical sociologists to treat NRTs only as closed 'black boxes' with a social impact. Conversely, practitioners of SSK interested in NRTs have had to follow the scientists down the corridors into the clinic to meet patients (Cussins 1996) as well as into politics (Mulkay 1997).

Given the lively cross-border trade just described, it is ironic that there are no contributions directly discussing gender and medical science and technology in this monograph. This was not planned, but the outcome of the vicissitudes of compiling monographs such as this. My only consolation is that work on gender and medical science is so much more developed than most of the areas covered by contributors to this volume that it has less need of this particular platform.

The Monograph and its Contents

In the original calls for contributions, two broad aims for this monograph were proposed: to publish papers which bridge the subfields of sociology of science and of medicine and to increase dialogue between sociologists researching science and technology related to health from different perspectives. Most of the recently published examples of rapprochement between the two subfields are applications of constructivist perspectives (*e.g.* Berg and Mol 1997, Casper and Berg 1995). For this monograph there was, deliberately, no prior commitment to any particular theoretical or methodological stance. The result of these calls is a mixture of perspectives, with a preponderance of broadly constructivist perspectives but also some notes of dissent, particularly in Abraham's chapter.

To encourage connections between chapters, contributors were asked to consider some or all of the following themes in relation to their particular topic: the extent of common concerns, contrasts and gaps in the relevant medical sociology and sociology of science literature; the relationships between different forms of knowledge identified in the research; the implications for general claims about the relationship between medical science and practice or medical and scientific authority in contemporary society and how these might be changing; and the implications for policy and future sociological research agendas.

The individual contributions have been arranged into three thematically organised sections, although, as will be clear to the reader, there are overlaps and common themes between chapters in different sections. Each section contains chapters deploying different conceptual approaches to indicate the diversity of perspectives applicable in the sociology of medical science and technology. The three chapters in the first section all address aspects of the connections between medical research and clinical practice or how they mutually constitute each other. The two chapters in the second section focus on the construction of professional tools, on opening up 'black boxes'. In the final section, two chapters examine the assessment and evaluation of medical technologies, taking us into the 'laboratory' of the NHS and also into national and transnational regulatory systems and health policy.

Science and Clinical Practice

In the first chapter, Nukaga and Cambrosio draw on ANT to examine the construction of genetic 'pedigrees' in genetic counselling clinics, that is the family trees that display connections between kin and the distribution of medical conditions among them. The genetic counselling clinic is often simultaneously a site for the collection of research data about 'family diseases' and a site for health care work with individuals who may be affected by such diseases. In providing a detailed ethnography of a mundane, taken-for-granted tool of inscription, Nukaga and Cambrosio's contribution is characteristic of much current work in SSK. They show how, notwithstanding the proliferation of complex new technologies in the 'new genetics', the family pedigree remains the basic technique for the clinical geneticist and constitutes an essential link between clinic and laboratory. The comparisons Nukaga and Cambrosio make between Japanese and Canadian practices indicate how the form and stability of family structures may affect what can be readily elicited from individuals about their kin and subsequently compiled into 'large pedigrees' for research purposes. But, in accordance with much recent SSK work, the authors argue pedigrees cannot be reduced to social relations. The pedigrees themselves contribute to new understandings and renderings of kinship, for example, posing new questions about respecting confidentiality. As Nukaga and Cambrosio demonstrate, medical pedigrees are means of co-ordinating the work of various kinds of health care professionals: genetic counsellors (who are themselves a heterogeneous group), diagnostic laboratory supervisors, clinical specialists and research geneticists. There is, in short, a complex division of labour which is, in part, hierarchically ordered. 'Routine work' tends to be delegated downwards away from 'researchers'.

Mueller's chapter focuses specifically on the division of labour in another setting where the boundaries between clinical setting and laboratory research are blurred, the academic medical centre dedicated to clinical trials of newly developed therapies. In her study of a US federally funded centre for AIDS therapy trials, Mueller documents the tensions between medical researchers and the nurses responsible for day-to-day management. The senior medical investigators directing the trials were, for the most part, away from the clinic, engaged in scientific entrepreneurship to recruit and retain financial resources for trials (*cf.* Latour 1987). The nurse trial-coordinators were responsible for recruiting and retaining human resources, patient-volunteers, for trials. Working with patient-volunteers on a daily basis, these nurses were unable to segment their role as caregiver from that of researcher. Some aspects of the tension between 'science' and 'care' that Mueller describes may be specific to the large and highly politically visible AIDS research programme or to the US health care system, for example, the problems of patient-volunteers without health insurance. But the general issues her chapter raises about the increasing segmentation of clinical research from clinical care and the emerging role of nurses in medical research may have wider application. Mueller rightly eschews a simple explanation of the nurse-doctor conflict in terms of participants' gender role socialisation. But her chapter raises questions about the extent to which supposedly 'feminine' skills have been tacitly built into research nurses' work (Davies 1995).

The inclusion of the chapter by Busby, Williams and Rogers on lay and biomedical understanding of musculoskeletal disorders in a section entitled 'Science and Clinical Practice' might seem paradoxical. But their work shows not so much the permeable boundaries between research laboratory and clinical sites as the gaps that exist between research and practice and the problems of moving between the 'laboratory' and routine patient care. As Star has pointed out, 'the links between science, medicine, and technology in work practices *themselves* are not transparent' (1995: 502 emphasis original). And it may be patients and primary health care professionals who have to do much of the articulation work. Busby *et al.* report a 'vacuum of understanding' between the 'high science' of rheumatology research and the development of strategies for disease management and patient care in the clinic. Drawing on work in both the sociology of science and of medicine, their analysis shows how the contrast between 'lay' and 'expert' understanding should not be drawn too sharply. Expert knowledge is not evenly distributed among medical professionals. In the UK, general medical practitioners (GPs) in particular are faced with the task of managing their relationships with patients in the absence of effective treatment while maintaining their professional identity as experts (*cf.* Singleton and Michael (1993).

Constructing Professional Tools

The two chapters in the second section both adapt established approaches in the sociology of scientific knowledge for use in case studies of the development of medical knowledge and, hence, of tools for use by health professionals. Atkinson, Batchelor and Parsons take us into the medical genetics research laboratories responsible for a 'major scientific breakthrough'. Their chapter is a narative analysis of scientists' retrospective accounts of the 'discovery' of the gene responsible for the most common form of adult muscular dystrophy. Since 1991, Myotonic Dystrophy has no longer been a disease of 'unknown' (but assumed genetic) causation. It is now attributable to an unstable sequence of DNA located within a specific gene, a reformulation which has implications for future research, treatments and generations. A diagnostic test for clinic use, a gene probe which can provide presymptomatic diagnosis of those whose family history indicates they may be at risk of Myotonic Dystrophy, has been developed.

Through analysing scientists' retrospective accounts of the processes and outcomes that led to their successful discovery claim, Atkinson *et al.* show how scientists employ different narrative registers simultaneously to account for their success (*cf.* Gilbert and Mulkay 1984). The scientists stressed both the predictability of the eventual outcome as the product of normal science and the unpredictability of the precise timing and place of the discovery and of the precise genetic anomaly involved. Atkinson *et al.* argue that sociologists of medicine should not take for granted the discovery claims and routine diagnostic knowledge that emerge from laboratories. Rather they should recognise the ways in which the priorities, opportunities and limits of clinical practice may be set by the discoveries and technologies emerging from laboratories, particularly in rapidly developing research fields such as molecular genetics.

The next chapter, by Bartley, Davey Smith and Blane, is not concerned with the 'cutting edge' of modern biomedical science. On the contrary, they examine the history of two statistical tools now routinely used in epidemiology and public health in explaining observed distributions of death or disease and in formulating preventive health policies. Arguably, preventive health policies should start from the 'facts' about disease patterns. But what the 'facts' are taken to be depends, in part, on the statistical measures used to translate the myriad instances of ill-health or death into manageable higher order inscriptions. Bartley *et al.*'s study draws on Latour's (1987) translation approach. They ask how it was that the Standardised Mortality Ratio (SMR) and the Odds Ratio came to be accepted as obligatory passage points, as indispensable transformations through which epidemiological data are expected to pass. They present some preliminary answers, based on their reading of secondary accounts of the history of public health, statistics and biographies of the key figures involved.

In the case of the SMR, their analysis shows that although a robust method for measuring life expectancy was widely sought from the mid-nineteenth century by social reformers and life insurance companies in Britain, the routine adoption of the SMR, a measure that took account of the age distribution of the populations under study, came about only slowly, many years after it was first proposed. Bartley *et al.* trace the complex and heterogeneous alliances whose concerns about recording the vital matters of individuals' life and death were gradually tied together in the work of the General Record Office. It was, they suggest, the successful deployment of age standardisation in particular public health controversies that established the SMR as the statistical device used almost without question today. Similarly, their historical account suggests that it was the alliances that a particular relative risk calculation served rather than the undisputed technical superiority of the method that explains why Odds Ratio comparisons have become accepted as the standard means of determining the relative importance of different putative causal factors in epidemiological studies.

Assessing and Regulating Medical Technologies

The two chapters in the third and final section of the book are both concerned with the assessment and control of the quality of health care, particularly in relation to technologies, and with state involvement in such evaluation. Abraham presents an analysis of the science and politics of regulating medicines, starting with an avowedly realist account of the scientific uncertainty that dogs the safety assessment of new medicines. Such uncertainty, he argues, provides the space in which social and political factors can enter into regulation '*under the guise of technical problem-solving* with relative ease' (Abraham, this volume, emphasis original). But the particular social and political factors at issue may vary according to national, supranational and international context. Abraham compares the development and operation of the British and US regulatory systems, concluding that the former has been more characterised by corporate bias prioritising industrial interests over consumers' interests. He argues that, in the United States, attempts by industrial interests to 'capture' regulatory authorities have been more visibly and effectively resisted. Abraham suggests that the emerging European Union framework for medicines' control is closer to the more technocratic British approach than the American one. His analysis has implications for our understanding of the current state and future of expert medical authority. On the one hand, as Gabe and Bury (1996) have argued in the case of *Halcion*, when controversy over the safety of individual drugs enters the public domain, medical authority can be seen to be fracturing and the privileged status of medical experts is demonstrably challenged by lay interests. But, simultaneously, Abraham suggests, the emerging global regulatory frameworks are likely to promote convergence of medical and scientific expertise and consensus in decision-making over drug licensing. This decision-making may be becoming more likely to occur behind doors which are relatively closed to consumer interest groups.

Faulkner's chapter examines the relatively new, rapidly growing and largely state-sponsored field of health services research, particularly health technology assessment (HTA) using the UK as a case study. He suggests that the UK's National Health Service can be conceptualised as a 'massive laboratory' for the construction of knowledge for HTA, analogous to the laboratories studied by sociologists of science and technology testing (*e.g.* Knorr-Cetina 1995, MacKenzie 1993). Observations produced under test conditions, as in random controlled clinical trials, are projected by interpretive techniques as predictions of how a particular technology (the term has a very wide application in HTA) might perform in the 'real world' of health care practice. These activities bring some 'strange bedfellows' together, including clinicians, epidemiologists, statisticians, health economists and sociologists. Faulkner examines the emerging alliances between these disciplines through analysis of their rhetoric about and for HTA. These rhetorics constitute professional 'boundary work' (Gieryn 1983, 1995), laying claims to professional territory by defining appropriate methods, concepts and agendas for HTA. Faulkner notes the paradox in this process. Distrust of traditional medical authority, as vested in clinicians, and of the efficacy and efficiency of many widely-used health technologies, is a core part of the case put forward for increased HTA. But HTA advocates are inviting us to place our trust in their emergent scientific expertise.

These seven chapters, ranging, as they do, from detailed ethnographies of the mundane genetics pedigree to the examination of transnational systems for regulating medicines illuminate many aspects of modern medical science and technology. They illustrate some of the many ways in which the study of medical science and technology can contribute to two independently flourishing subfields of sociology and promote the forging of further links between them. And they indicate the significance of medical science and technology in contemporary society and the relevance of studying them for some contemporary debates in social theory. Running through much of the work in the sociology of medicine and of science since the 1970s described in this Introduction has been the view that medical and scientific power and authority are being subjected to sustained challenge (*e.g.* Gabe *et al.* 1994, Irwin and Wynne 1996a). Indeed, whether intentionally or not, this critical sociology has become part of this challenge, exemplifying the reflexivity of late modernity (Giddens 1990, Elston 1994). According to some, the extent of challenge indicates the onset of a post-modern era in which modernist science carries no special authority. But the contributions to this volume suggest there is still much room for debate about whether medical and scientific authority is being displaced. New medical technologies have been strongly implicated in the post-modern vision of the flexible, uncertain body and often seen as threatening to social order. But before accepting the romantic vision of the cyborg culture or the doom-laden scenario as imminent, close examination of these new (and old) medical technologies and the practices that surround them, is surely appropriate.

Acknowledgements

I am grateful to Jonathan Gabe for his comments on drafts of this introduction. Without the support of grants from the Economic and Social Research Council to study 'the changing organization of animal experimentation (Y305253003) and 'the public controversy over animal experimentation since 1960' (R000234514) I might never have had the opportunity to venture into sociology of science territory and hence to reflect on my own professional culture, medical sociology.

Notes

1 Bynum (1994), Lawrence (1985, 1994) and Warner (1985) provide useful historical overviews.
2 'Sociology of medicine' (or 'medical sociology') is used inclusively in this chapter to refer to the sociological study of all aspects of health and health care and, as is generally evident from the context, not just the work of 'medics'. 'Sociology of science' is also used inclusively to cover more than work in the currently dominant paradigm, sociology of scientific knowledge (SSK). On the indexicality of terms such as 'science' and 'technology' see Pinch *et al.* (1992).
3 That there had been several studies of unorthodox medical knowledge (*e.g.* Wallis 1979) supports Richards' claim about 'orthodox' medicine (see also Saks 1995). One significant exception to Richards' charge, Latour's study of *The Pasteurization of France*, (of how 'Pasteurism' became orthodoxy) first published in French in 1984, became available in English in 1988.
4 A 'black box' is a symbol used by cyberneticians and engineers to represent a piece of machinery or set of commands for which, for the purposes in-hand, they need only to consider inputs and outputs and not its internal workings (Bijker *et al.* 1987).
5 For fuller accounts of the development of medical sociology see, for example, Bury (1997) or Gerhardt (1989). On the sociology of science and technology see, for example, Jasanoff et al (1995), Webster (1991) or Yearley (1988).
6 Cyborgs are hybrids between cybernetic devices and organisms but Haraway uses the term metaphorically to refer to other kinds of hybrid 'beings'. Latour (1993) and Michael (1996), writing from an ANT perspective, argue that hybrids are not peculiarly post-modern.

References

Abbott, A. (1988) *The System of Professions.* Chicago: University of Chicago Press.
Abraham, J. (1995) *Science, Politics and the Pharmaceutical Industry*. London: UCL Press.
Annandale, E. and Clark, J. (1996) What is gender? Feminist theory and the sociology of human reproduction, *Sociology of Health and Illness*, 18, 17–44.

Arksey, H. (1994) Expert and lay participation in the construction of medical knowledge, *Sociology of Health and Illness*, 16, 448–69.
Armstrong, D. (1983) *Political Anatomy of the Body: Medical Knowledge in Britain in the Twentieth Century*. Cambridge: Cambridge University Press.
Ashmore, M., Mulkay, M. and Pinch, T. (1989) *Health and Efficiency: A Sociology of Health Economics*. Milton Keynes: Open University Press.
Atkinson, P. (1995) *Medical Talk and Medical Work*. Thousand Oaks, CA and London: Sage.
Atkinson, P. (1997) Anselm Strauss: an appreciation, *Sociology of Health and Illness*, 19, 367–72.
Atkinson, P. and Delamont, S. (1990) Professions and powerlessness, *Sociological Review*, 38, 90–110.
Atkinson, P. and Parsons, E. (1992) Lay construction of genetic risk, *Sociology of Health and Illness*, 14, 437–55.
Balmer, B. (1996) Managing mapping in the Human Genome Project, *Social Studies of Science*, 26, 531–73.
Barber, B. (1967) *Drugs and Society*. New York: Russell Sage Foundation.
Barber, B. (1990) *Social Studies of Science*. New Brunswick: Transaction Publishers.
Bartley, M. (1990) Do we need a strong programme in medical sociology? *Sociology of Health and Illness*, 12, 371–90.
Berg, M. (1992) The construction of medical disposals. Medical sociology and medical problem-solving in clinical practice, *Sociology of Health and Illness*, 14, 151–80.
Berg, M. (1995) Turning a practice into a science: reconceptualising postwar medical practice, *Social Studies of Science*, 25, 437–76.
Berg, M. (1997a) Problems and promises of the protocol, *Social Science and Medicine*, 44, 1081–8.
Berg, M. (1997b) *Rationalizing Medical Work: Decision-Support Techniques and Medical Practice*. Cambridge, MA: MIT Press.
Berg, M. and Mol, A. (eds) (1997) *Differences in Medicine: Unravelling Practices, Techniques and Bodies*. Durham, N.C.: Duke University Press.
Bijker, W. (1993) Do not despair: there is life after constructivism, *Science, Technology and Human Values*, 18, 113–38.
Bijker, W., Hughes, T. and Pinch, T. (eds) (1987) *The Social Construction of Technological Systems*. Cambridge, MA: MIT Press.
Bijker, W. W. and Law, J. (eds) (1992) *Shaping Technology/Building Society*. Cambridge, MA: MIT Press.
Bloor, D. (1976/1991) *Knowledge and Social Imagery*. 1st edition London: Routledge and Kegan Paul. 2nd edition Chicago: University of Chicago Press.
Bloor, M. (1995) *The Sociology of HIV Transmission*. London and Thousand Oaks, CA: Sage.
Blume, S. (1991) *Insight and Industry: On the Dynamics of Technological Change in Medicine*. Cambridge, MA: MIT Press.
Bury, M. (1986) Social constructionism and the development of medical sociology, *Sociology of Health and Illness*, 8, 137–69.
Bury, M. (1997) *Health and Illness in a Changing Society*. London: Routledge.
Bynum, W.F. (1994) *Science and the Practice of Medicine in the Nineteenth Century*. Cambridge: Cambridge University Press.
Callon, M. (1986) The sociology of an actor-network: the case of the electric vehicle.

In Callon, M., Law, J. and Rip, A. (eds) *Mapping the Dynamics of Science and Technology*. Basingstoke: Macmillan.
Casper, M.J. and Berg, M. (1995) Constructivist perspectives on medical work: medical practices and science and technology studies, *Science, Technology and Human Values*, 20, 395–407.
Clarke, A.E. and Montini, T. (1993) The many faces of RU486: tales of situated knowledges and technological contestation, *Science, Technology and Human Values*, 18, 42–78.
Collins, A., Kendall, G. and Michael, M. (1998 f.c.) Resisting a diagnostic technique: the case of reflex anal dilation, *Sociology of Health and Illness*, 19.
Collins, H.M. (1985) *Changing Order: Replication and Induction in Scientific Practice*. Revised edition. Chicago: University of Chicago Press.
Conrad, P. (1997) Public eyes and private genes: historical frames, news constructions, and social problems, *Social Problems*, 44, 139–54.
Cussins, C. (1996) Ontological choreography: agency through objectification in infertility clinics, *Social Studies of Science*, 26, 575–610.
Davies, C. (1995) *Gender and the Professional Predicament of Nursing*. Buckingham: Open University Press.
Davis, P. (ed) (1996) *Contested Ground: Public Purpose and Private Interest in the Regulation of Prescription Drugs*. New York: Oxford University Press.
Davis, P. (1997) *Managing Medicines: Public Policy and Therapeutic Drugs*. Buckingham: Open University Press.
Delamont, S. (1987) Three blind spots? A comment on the sociology of science by a puzzled outsider, *Social Studies of Science*, 17, 163–70.
Doyal, L. with Pennell, I. (1979) *The Political Economy of Health*. London: Pluto Press.
Draper, E. (1991) *Risky Business: Genetic Testing and Exclusionary Practice in Hazardous Industries*. Cambridge: Cambridge University Press.
Elston, M.A. (1993) Women doctors in a changing profession: the case of Britain. In Riska, E. and Weger, K. (eds) *Gender, Work and Medicine: Women and the Medical Division of Labour*. London and Thousand Oaks, CA: Sage.
Elston, M.A. (1994) The anti-vivisectionist movement and the science of medicine. In Gabe, J. Kelleher, D. and Williams, G. (eds) *Challenging Medicine*. London: Routledge.
Epstein, S. (1995) The construction of lay expertise: AIDS activism, and the forging of credibility in the reform of clinical trials, *Science, Technology and Human Values*, 20, 408–437.
Fleck, L. (1935/1979) *Genesis and Development of a Scientific Fact*. Chicago: University of Chicago Press.
Foucault, M. (1976) *The Birth of the Clinic: An Archaeology of Medical Perception*. London: Tavistock.
Fox, R. (1959/1974) *Experiment Perilous*. Philadelphia: University of Pennsylvania Press.
Fox, R. (1980) The evolution of medical uncertainty, *Milbank Memorial Fund Quarterly*, 58, 1–49.
Fox, R. and Swazey, J.P. (1974) *The Courage to Fail: A Social View of Organ Transplantation and Dialysis*. Chicago: Chicago University Press.
Fujimura, J. (1996) *Crafting Science: A Sociohistory of the Quest for the Genetics of Cancer*. Cambridge, MA: Harvard University Press.

Fujimura, J. and Chou, D.Y. (1994) Dissent in science: styles of scientific practice and the controversy over the cause of AIDS, *Social Science and Medicine*, 33, 226–51.

Gabe, J. (ed) (1995) *Medicine, Health and Risk: Sociological Approaches*. Sociology of Health and Illness Monograph 1. Oxford: Blackwell.

Gabe, J. and Bury, M. (1996) Halcion nights: a sociological account of medical controversy, *Sociology*, 30, 447–69.

Gabe, J., Kelleher, D. and Williams, G. (eds) (1994) *Challenging Medicine*. London: Routledge.

Gerhardt, U. (1989) *Ideas about Illness: An Intellectual and Political History of Medical Sociology*. London: Macmillan.

Gibbons, M., Limoges, C., Nowotny, H., Schwartzman, S., Scott, P. and Trow, M. (1994) *The New Production of Knowledge*. London and Thousand Oaks, CA: Sage.

Giddens, A. (1990) *The Consequences of Modernity*. Cambridge: Polity.

Gieryn, T.F. (1983) Boundary-work and the demarcation of science from non-science: strains and interests in professional ideologies of science, *American Sociological Review*, 48, 781–95.

Gieryn, T.F. (1995) Boundaries of science. In Jasanoff, S., Markle, G.E., Petersen, J.C. and Pinch, T. (eds) *Handbook of Science and Technology Studies*. Thousand Oaks, CA and London: Sage.

Gilbert, N. and Mulkay, M. (1984) *Opening Pandora's Box: A Sociological Analysis of Scientists' Discourse*. Cambridge: Cambridge University Press.

Good, B.J. (1994) *Medicine, Rationality and Experience: An Anthropological Perspective*. Cambridge: Cambridge University Press.

Good, M-J. (1995) Cultural studies of biomedicine: an agenda for research, *Social Science and Medicine*, 41, 461–73.

Gray, C.H. (1995) *The Cyborg Handbook*. London: Routledge.

Grinyer, A. (1995) Risk, the real world and naive sociology. In Gabe, J. (ed) *Medicine, Health and Risk: Sociological Approaches. Sociology of Health and Illness Monograph 1.* Oxford: Blackwell.

Hafferty, F.W. and Light, D.W. (1995) Professional dynamics and the changing nature of medical work, *Journal of Health and Social Behavior* (Extra Issue), 132–53.

Ham, C., Robinson, R. and Benzeval, M. (1990) *Health Check: Health Care Reforms in an International Perspective*. London: King's Fund.

Haraway, D. (1989) *Primate Visions*. London: Routledge.

Haraway, D. (1991) *Simians, Cyborgs, and Women*. London: Free Association Books.

Hart, G., Boulton, M., Fitzpatrick, R., McLean, J. and Dawson, J. (1992) 'Relapse' to unsafe sexual behaviour amongst gay men: a critique of recent behavioural HIV/AIDS research, *Sociology of Health and Illness*, 14, 216–32.

Hartland, J. (1996) Automating blood pressure measurements: the division of labour and the transformation of method, *Social Studies of Science*, 26, 71–94.

Hilgartner, S. (1995) The human genome project. In Jasanoff, S., Markle, G.E., Petersen, J. and Pinch, T. (eds) *Handbook of Science and Technology Studies*. London and Thousand Oaks, CA: Sage.

Hirschauer, S. (1991) The manufacture of bodies in surgery, *Social Studies in Science*, 21, 279–319.

Hollis, M. and Lukes, S. (eds) (1982) *Rationality and Relativism*. Oxford: Blackwell.

Hubbard, R. (1990) *The Politics of Women's Biology*. New Brunswick and London: Rutgers University Press.

Indyk, D. and Rier, D.A. (1993) Grassroots AIDS knowledge: implications for the boundaries of science and collective action, *Knowledge: Creation, Diffusion, Utilization*, 15, 3–43.

Irwin, A. and Wynne, B. (eds) (1996a) *Misunderstanding Science? The Public Reconstruction of Science and Technology*. Cambridge: Cambridge University Press.

Irwin, A. and Wynne, B. (1996b) Introduction. In Irwin, A. and Wynne, B. (eds) *Misunderstanding Science? The Public Reconstruction of Science and Technology*. Cambridge: Cambridge University Press.

Jasanoff, S. (1990) *The Fifth Branch: Science Advisers as Policymakers*. Cambridge, MA: Harvard University Press.

Jasanoff, S., Markle, G.E., Petersen, J.C. and Pinch, T. (eds) (1995) *Handbook of Science and Technology Studies*. Thousand Oaks, CA and London: Sage.

Jewson, N. (1976) The disappearance of the sick man from medical cosmology, 1770–1870, *Sociology*, 10, 225–44.

Jordanova, L. (1989) *Sexual Visions. Images of Gender in Science and Medicine between the Eighteenth and Twentieth Centuries*. Brighton: Harvester.

Keller, E. Fox (1995) The origin, history and politics of the subject called 'Gender and Science'. In Jasanoff, S., Markle, G.E., Petersen, J.C. and Pinch, T. (eds) (1995) *Handbook of Science and Technology Studies*. Thousand Oaks, CA and London: Sage.

Knorr-Cetina, K. (1995) Laboratory studies: the cultural approach to the study of science. In Jasanoff, S., Markle, G.E., Petersen, J.C. and Pinch, T. (eds) *Handbook of Science and Technology Studies*. Thousand Oaks, CA and London: Sage.

Kuhn, T. (1970) *The Structure of Scientific Revolutions*. 2nd edition. Chicago: University of Chicago Press.

Lambert, H. and Rose, H. (1996) Disembodied knowledge? Making sense of medical science. In Irwin, A. and Wynne, B. (eds) *Misunderstanding Science?* Cambridge: Cambridge University Press.

Latour, B. (1987) *Science in Action*. Milton Keynes: Open University Press.

Latour, B. (1988) *The Pasteurization of France*. Cambridge, MA: Harvard University Press. (Originally published in 1984 as *Les microbes: guerre et paix suivi de irréductions*. Paris: A.M. Metaille.)

Latour, B. (1993) *We Have Never Been Modern*. Hemel Hempstead: Harvester-Wheatsheaf.

Latour, B. and Woolgar, S. (1986) *Laboratory Life*. 2nd edition. Princeton, NJ: Princeton University Press.

Law, J. (1992) Notes on the theory of the actor-network: ordering, strategy, and heterogeneity, *Systems Practice*, 5, 379–93.

Lawrence, C. (1985) Incommunicable knowledge: Science, technology and the clinical art in Britain, 1850–1914, *Journal of Contemporary History*, XX, 503–20.

Lawrence, C. (1994) *Medicine in the Making of Modern Britain: 1700–1920*. London. Routledge.

Levidow, L. and Young, B. (eds) (1982) *Science, Technology and the Labour Process*. Vol 2. London: Humanities Press.

Lippman, A. (1991) Prenatal genetic testing and screening: constructing needs and reinforcing inequities, *American Journal of Law and Medicine*, 17, 15–20.

Löwy, I. (ed) (1993) *Medicine and Change: Historical and Sociological Studies of Medical Innovation*. Paris: INSERM.

Lynch, M. (1985) *Art and Artifact in Laboratory Science: A Study of Shop Work and Shop Talk in a Research Laboratory*. London: Routledge.
Macdonald, K.M. (1995) *The Sociology of the Professions*. London and Thousand Oaks, CA: Sage.
MacKenzie, D. (1993) *Inventing Accuracy: A Historical Sociology of Nuclear Missile Guidance*. Cambridge, MA: MIT Press.
MacKenzie, D. and Wajcman, J. (eds) (1985) *The Social Shaping of Technology: How the Refrigerator got its Hum*. Milton Keynes: Open University Press.
McKinlay, J. (1981) From 'promising report' to 'standard procedure': seven stages in the career of a medical innovation, *Milbank Memorial Fund Quarterly: Health and Society*, 59, 374–411.
McNeil, M., Varcoe, I. and Yearley, S. (eds) (1990) *The New Reproductive Technologies*. London: Macmillan.
Marteau, T. and Richards, M.R. (eds) (1996) *The Troubled Helix: Social and Psychological Implications of the New Human Genetics*. Cambridge: Cambridge University Press.
Martin, B. (1997) Sticking a needle into science: the case of polio vaccines and the origin of AIDS, *Social Studies of Science*, 26, 245–76.
Martin, E. (1989) *The Woman in the Body: A Cultural Analysis of Reproduction*. Milton Keynes: Open University Press.
Martin, E. (1994) *Flexible Bodies*. Boston: Beacon Books.
Maynard, A. and Chalmers, I. (eds) (1997) *Non-Random Reflections on Health Services Research*. London: BMJ Publishing Group.
Melia, K. (1987) *Learning and Working: The Occupational Socialization of Nurses*. London: Tavistock.
Merton, R., Reader, G. and Kendall, P.L. (eds) (1957) *The Student-Physician: Introductory Studies in the Sociology of Medical Education*. Cambridge, MA: Harvard University Press.
Michael, M. (1996) *Constructing Identities*. Thousand Oaks, CA and London: Sage.
Mol, A. and Elsman, B. (1996) Detecting disease and designing treatment. Duplex and the diagnosis of diseased leg vessels, *Sociology of Health and Illness*, 18, 609–31.
Montini, T. and Ruzek, S. (1993) Overturning orthodoxy: the emergence of breast cancer treatment policy, *Research in the Sociology of Health Care*, 8, 3–32.
Mulkay, M. (1997) *The Embryo Research Debate: Science and the Politics of Reproduction*. Cambridge: Cambridge University Press.
Nelkin, D. and Lindee, S. (1995) *The DNA Mystique: The Gene as a Cultural Icon*. New York: Freeman and Co.
Nettleton, S. (1995) *The Sociology of Health and Illness*. Cambridge: Polity.
Nicolson, M. and McLaughlin, C. (1987) Social constructionism and medical sociology: a reply to M.R. Bury, *Sociology of Health and Illness*, 9, 107–26.
Nicolson, M. and McLaughlin, C. (1988) Social constructionism and medical sociology: a study of the vascular theory of multiple sclerosis, *Sociology of Health and Illness*, 10, 234–61.
Oudshoorn, N. (1994) *Beyond the Natural Body: An Archaeology of Sex Hormones*. London: Routledge.
Petersen, J.C. and Markle, G.E. (1981) Expansion of conflict in cancer controversies.

In Kriesberg, L. (ed) *Research in Social Movements, Conflict and Change*, Vol. 4. Greenwich, GT: JAI Press.

Pfeffer, N. (1993) *The Stork and the Syringe: A Political History of Reproductive Medicine*. Cambridge: Polity.

Pickstone, J.V. (ed) (1992) *Medical Innovations in Historical Perspective*. Basingstoke: Macmillan.

Pickstone, J.V. (1993) Ways of knowing – towards a historical sociology of science, technology and medicine, *British Journal for the History of Science*, 26, 433–58.

Pinch, T., Ashmore, M. and Mulkay, M. (1992) Technology, testing, text: clinical budgeting in the U.K. National Health Service. In Bijker, W.E. and Law, J. (eds) *Shaping Technology/Building Society*. Cambridge, MA: MIT Press.

Press, N. and Browner, C.H. (1997) Why women say yes to prenatal diagnosis, *Social Science and Medicine*, 45, 979–90.

Prout, A. (1996) Actor-network theory and medical sociology: an illustrative analysis of the metered dose inhaler, *Sociology of Health and Illness*, 18, 198–219.

Richards, E. (1988) The politics of therapeutic evaluation: the vitamin C and cancer controversy, *Social Studies of Science*, 18, 653–701.

Richads, E. (1991) *Vitamin C and Cancer: Medicine or Politics?* London: Macmillan.

Richards, M.R. (1993) The new genetics: some issues for social scientists, *Sociology of Health and Illness*, 15, 567–86.

Riska, E. and Wegar, K. (eds) (1993) *Gender, Work and Medicine: Women and the Medical Division of Labour*. London and Thousand Oaks, CA: Sage.

Rose, H. (1994) *Love, Power and Knowledge*. Cambridge: Polity Press.

Saks, M. (1995) *Professions and the Public Interest: Medical Power, Altruism and Alternative Medicine*. London: Routledge.

Schiebinger, L. (1989) *The Mind has No Sex? Women in the Origins of Modern Science*. Cambridge, MA: Harvard University Press.

Schroeder, R. (1996) *Possible Worlds: The Social Dynamic of Virtual Reality Technology*. Boulder, CO: Westview Press.

Searle, J. (1995) *The Construction of Social Reality*. London: Allen Lane.

Shapin, S. (1989) The invisible technician, *American Scientists*, 77, 554–63.

Shilling, C. (1993) *The Body in Social Theory*. Thousand Oaks, CA and London: Sage.

Singleton, V. and Michael, M. (1993) Actor-networks and ambivalence: general practitioners in the cervical screening programme, *Social Studies of Science*, 23, 227–64.

Stacey, M. (ed) (1992) *Changing Human Reproduction: Social Science Perspectives*. London and Thousand Oaks, CA: Sage.

Stanworth, M. (ed) (1987) *Reproductive Technologies: Gender, Motherhood and Medicine*. Cambridge: Polity.

Star, S. Leigh (1995) Epilogue: work and practice in social studies of science, medicine and technology, *Science, Technology and Human Values*, 20, 501–7.

Storer, N.W. (1971) Introduction to Merton, R.K. *The Sociology of Science: Theoretical and Empirical Investigations*. Chicago: University of Chicago Press.

Strauss, A., Fagerhaugh, S., Suczek, B. and Wiener, C. (1985) *Social Organization of Medical Work*. Chicago: University of Chicago Press.

Strong, P. (1979) Sociological imperialism and the profession of medicine, *Social Science and medicine*, 13A, 199–215.

Timmermans, (1996) Saving lives or saving multiple identities? The double dynamic of resuscitation scripts, *Social Studies of Science*, 26, 767–97.

Traweek, S. (1993) An introduction to cultural and social studies of science and technologies, *Culture, Medicine and Psychiatry*, 17, 3–25.

Wallis, R. (ed) (1979) *On the Margins of Science: The Social Construction of Rejected Knowledge*. Sociological Review Monograph. Keele: University of Keele.

Warner, J. Harley (1985) Science in medicine, *Osiris*, 2nd series, 1, 37–58.

Webster, A. (1991) *Science, Technology and Society*. Basingstoke: Macmillan.

Webster, A. (1994) University-corporate ties and the construction of research agendas, *Sociology*, 28, 123–42.

Williams, G. and Popay, J. (1994) Lay knowledge and the privilege of experience. In Gabe, J., Kelleher, D. and Williams, G. (eds) *Challenging Medicine*. London: Routledge.

Williams, S.J. (1997) Modern medicine and the 'uncertain body': from corporeality to hyperreality?, *Social Science and Medicine*, 45, 1041–9.

Williams, S.J. and Calnan, M. (eds) (1996) *Modern Medicine: Lay Perspectives and Experiences*. London: UCL Press.

Willis, E. (1994) The social relations of HIV testing technology. In Scott, S. and Williams, G. (eds) *Private Risks and Public Dangers*. Aldershot: Avebury.

Witz, A. (1994) The challenge of nursing. In Gabe, J., Kelleher, D. and Williams, G. (eds) *Challenging Medicine*. London: Routledge.

Wolpert, L. (1992) *The Unnatural Nature of Science: Why Science Does Not Make (Common) Sense*. London: Faber and Faber.

Woolgar, S. (ed) (1988) *Knowledge and Reflexivity: New Frontiers in the Sociology of Knowledge*. London and Thousand Oaks, CA: Sage.

Wright, P. and Treacher, A. (eds) (1982) *The Problem of Medical Knowledge: Examining the Social Construction of Medicine*. Edinburgh: Edinburgh University Press.

Wynne, B. (1991) Knowledges in context, *Science, Technology and Human Values*, 16, 111–21.

Wynne, B. (1996) Misunderstood misunderstandings: social identities and public uptake of science. In Irwin, A. and Wynne, B. (eds) *Misunderstanding Science?* Cambridge: Cambridge University Press.

Yearley, S. (1988) *Science, Technology and Social Change*. London: Unwin Hyman.

Yoxen, E.J. (1987) Seeing with sound: a study of the development of medical images. In Bijker, W.E., Hughes, T.P. and Pinch, T. (eds) *The Social Construction of Technological Systems*. Cambridge, MA: MIT Press.

1. Medical pedigrees and the visual production of family disease in Canadian and Japanese genetic counselling practice

Yoshio Nukaga and Alberto Cambrosio

Introduction

At the end of the introduction to *Ethics and Human Genetics* (Wertz and Fletcher 1989: xxvii–xxix), J.C. Fletcher, a medical ethicist and a 'hearing son . . . of deaf parents', recalls his personal experience with genetic counselling: 'I and countless others were told that my father lost his hearing in 1904, when at age four, he was "struck by lightning" standing on the back porch of his Alabama farm home from watching a storm. . . . The story of my father's miraculous survival was literally a legend in his time'. By 1980, Fletcher's 'growing involvement with medical geneticists, unanswered questions about the actual cause of [his] parents' deafness, and three maturing children led [him] to seek help from genetic counsellors to assess the genetic risks of deafness in [his] family'. The story of his father's deafness was examined through discussions with neurologists and with his father's cousin, and Fletcher realised that the lightning story was false. The cousin recalled that the father had become 'very sick, was taken to the doctor for a long time, and came back deaf'. In the end, the exact cause of the father's deafness (meningitis? viral infection?) was not uncovered. Still, the outcome of this episode was that the old family story was examined, rejected and replaced by a medical account.

Fletcher's testimony can be used to illustrate the gap between, in Mishler's (1984) terms, the decontextualised 'voice of medicine' and the biographical narrative of the 'voice of the lifeworld'. Yet, such an interpretation is unsatisfactory since not only, as pointed out by Atkinson (1995), are there several voices of medicine, but, more importantly for our present purpose, it is not simply a question of replacing one story or 'voice' with another. In fact, enquiries into family stories such as Fletcher's routinely resort to the drawing of a medical pedigree, *i.e.* a 'family tree constructed from a person's knowledge of their family relations and the conditions each may or may not have had' (see Figure 1).[1] Indeed, 'even in the age of the new genetics', medical pedigrees still constitute the 'basic investigative tool' in the genetic clinic (Richards 1996: 250). Thus, medical genetic inquiries involve a series of translations from a web of oral narratives to a sequence of visual inscriptions which, in turn, become part of larger inscriptions connecting medical pedigrees to the visual display of, say, cytogenetic or molecular biological test results. To be sure, pedigrees also tell (different) stories, but to focus simply on their narrative dimension is to miss the opportunity to investigate exactly *how* medicine produces its objects, by analysing the sequential construction, display, combination and mobilisation of various representational tools (Berg 1996, Berg and Mol 1997, Howell 1995, Raffel 1979).

In his discussion of 18th century hospitals, Foucault (1977: 189–90) noted that the latter had already become 'great laboratories for scriptuary and documentary methods', a 'system of intense registration and documentary accumulation'. As a result, individuals were placed 'in a field of surveillance', situated 'in a network of writing' and engaged 'in a whole mass of documents that capture[d] and fix[ed] them'. In the case of genetic counselling, individuals are not only situated in a network of writing; they also become, literally, part of the medical pedigree's network of icons. In other words, they are translated into elements of a collective, familial, and thus biosocial body (Rabinow 1992). It is not simply that individuals are replaced in the 'context' of their families; rather, the redefinition of individuals through their association with genetically-based notions of risk leads to the simultaneous establishment of new notions of family and population (Richards 1996, Strathern 1992), thus modifying simultaneously the context and the relation of the individual to that context.

While functioning as tools for the production of medical evidence, that is, for the production of the family diseases of which they are allegedly a record (Raffel 1979), medical pedigrees also operate as 'boundary objects' (Star and Griesemer 1989) connecting both different professional practices as well as external and internal, collective and individual aspects of the body. Thus, their power as clinical tools lies less in their 'static' properties than in the fact that they can be mobilised as part of an expanding network of molecular evidence around which a practice such as genetic counselling is increasingly organised. The purpose of this paper is to shed some light on these processes and thus to constitute them into proper objects of inquiry.

Methodology

Most of the findings reported in this chapter originate in comparative, ethnographic fieldwork carried out from January 1994 to August 1995 in Canada and Japan. The decision to collect data in two widely-different national settings was strategically oriented to the task of exploring the taken-for-granted elements of genetic counselling practices. The fieldwork consisted of both interviews and participant-observation. The former were carried out, in Canada, with 29 clinical workers – 4 Ph.D. researchers, 8 MD genetic counsellors, 14 MS genetic counsellors, and 3 nurses – and, in

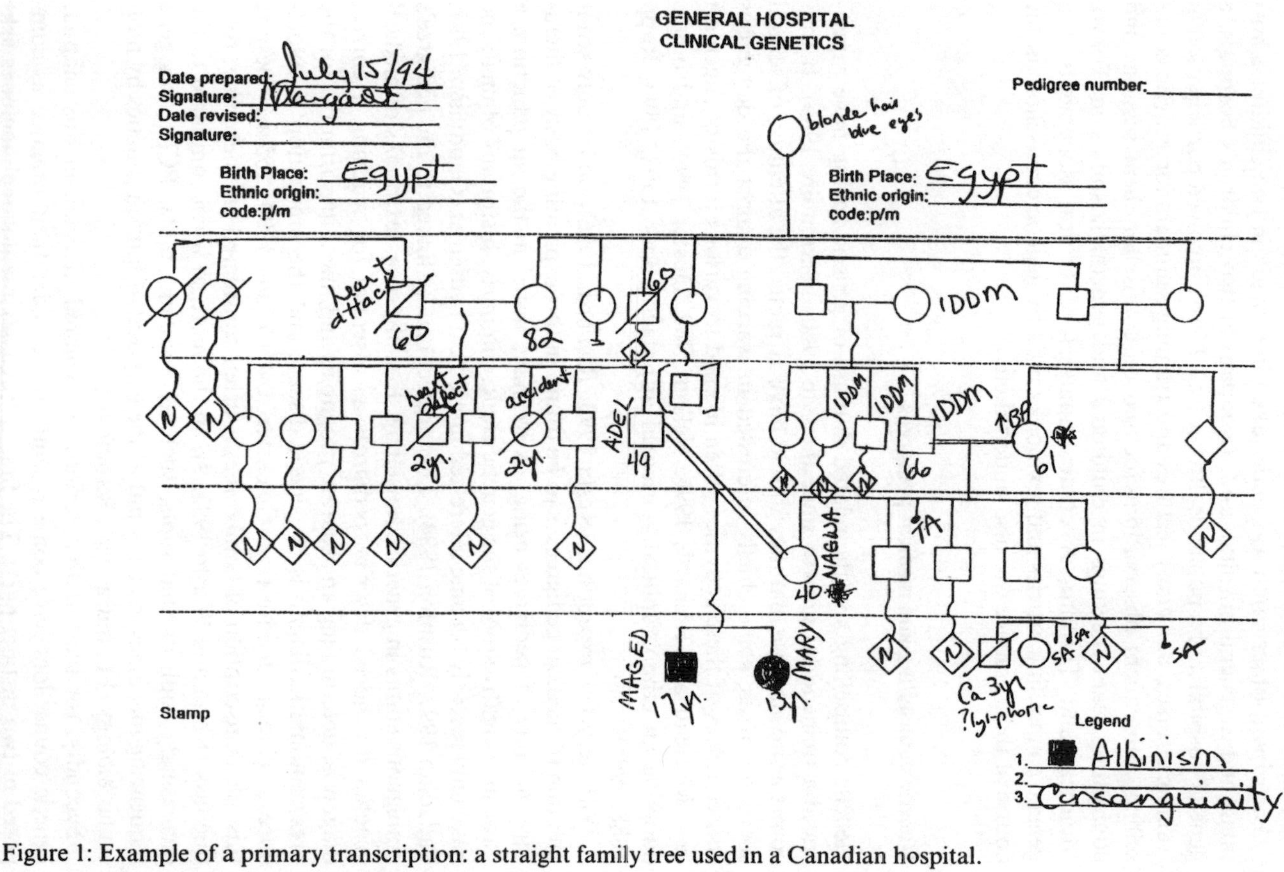

Figure 1: Example of a primary transcription: a straight family tree used in a Canadian hospital.

Japan, with 27 clinical workers – 3 Ph.D. researchers, 17 MD genetic counsellors, 7 public health nurses. Interviews were tape-recorded, except for informal discussions with laboratory workers, in which case detailed notes were made immediately after the discussion.

Participant-observation was undertaken in Canada in two children's hospitals and a general hospital, and, in Japan, in two children's hospitals, a general hospital and a public health centre.[2] It first involved participation in training courses, university lectures and training seminars for genetic counselling. Subsequent observations focused on counsellor-client sessions and counselling team meetings in children's and general hospitals in Tokyo, Montreal and Toronto. A content-analysis of clinical documents and genetic counselling articles and textbooks used by genetic counsellors in the course of their practice was also undertaken.

Genetic counselling and medical pedigrees

Genetic counselling and the related domain of genetic testing have experienced a tremendous growth since the late 1960s (Terrenoire 1986). It thus comes as no surprise that they should have attracted the attention of several social scientists, some chiefly interested in warning against the dangerous consequences of these activities, often likened to various forms of eugenism (*e.g.* Marteau and Richards 1996, Nelkin 1989, Duster 1990), and others adopting an ethnographically-oriented approach (Bosk 1992, 1993, Rapp 1988, 1995).

With very few exceptions (Resta 1993, Richards 1996), no or only scant mention of medical pedigrees is to be found in this expanding body of literature, in spite of pedigrees being, as we have noted in the introduction, a basic investigative tool of the genetic clinic. Similarly, historians of medicine have only recently become interested in medical pedigrees (Gaudillière 1997, Palladino 1997, Rushton 1994). This neglect can be linked to the pedigree's ambiguous status in genetic counselling. For instance, textbooks of medical genetics do indeed describe pedigrees as essential for reaching a correct diagnosis, determining an accurate prognosis and, most importantly, for the presymptomatic diagnosis of genetic disease and the prevention or avoidance of clinical disease (*e.g.* Gelehrter and Collins 1990: 255–6). Despite this, the transcription of family trees is often considered an unexciting, routine task (Resta and Wcislo 1993: 9), all the more so when compared to the increasingly sophisticated tools, such as Southern blots, PCR, and gene sequencing machines, recently put at the disposal of human genetics by molecular biology (Fujimura 1996, Rabinow 1996).

Secondly, for reasons that cannot be explored in detail in this chapter, genetic counsellors have come recently to consider their practice as composed of two distinct parts. The first, a 'preassessment stage', includes asking the 'reason for referral', taking a 'family history', collecting 'family history information', and practising a 'clinical examination and laboratory tests of relatives'. The second, a 'genetic counselling stage' or 'communication stage', deals with the 'nature and consequence of the disorder, recurrence risk, means of modification of consequences, means of prevention of recurrence (prenatal diagnosis and counselling)' (Thompson *et al.* 1991). This dichotomy relegates such practices as the taking of family pedigrees into the realm of the factual, unproblematic, routine activities in need of no further scrutiny.

And yet, as historians have noted, even such 'mundane' tools as medical files do not spring from the head of Zeus: it takes time and effort for them to reach their present unproblematic, routine status, thanks to processes that warrant close investigation (*e.g.* Howell 1995: 42–56). Moreover, it is our belief that the establishment of sharp distinctions between the various aspects of genetic counselling belies their intimate, mutually constitutive connections. We would go so far as to claim that it is precisely the ambiguous status of medical pedigrees – omnipresent, essential and yet taken for granted and almost invisible – that turns them into a sociologically-relevant topic. Medical pedigrees, then, can be understood as the visual tools or 'inscriptions' (Latour 1990) used by clinical and laboratory workers to make visible the invisible knowledge of the family and thus constitute the family and 'family disease' into an object of medical intervention.

The production of medical pedigrees

By following the genetic counsellors' work it is possible analytically to divide the process of visual documentation of family data into sequential elements: primary transcription, secondary transcriptions, combination and circulation. In spite of the obvious differences between laboratory and clinical work, these categories bear some resemblance to Amann and Knorr-Cetina's (1990) distinction between proto-data (*i.e.* uncertain, unstable, visually flexible documents), evidence (*i.e.* stabilised, published facts) and accepted theory, a distinction aimed at focusing attention on how proto-data are transformed into evidence through various forms of tinkering.

In the case of genetic counselling and medical pedigrees, the following processes can be distinguished. In *primary transcription*, family data are translated by genetic counsellors from oral narratives into a hand-written family tree characterised by the contrasting presence of icons depicting normality and abnormality. In primary transcription, counsellors focus on the *family* as the collective site of illness.

Secondary transcriptions within local clinical settings translate primary trees into medical evidence, *i.e.* into documents with both a clinical and a legal-administrative meaning, raising, for instance, the thorny issue of who owns this newly created 'family information'. Secondary transcriptions simultaneously standardise and stabilise the data produced by primary transcriptions, thus acting as a condition of possibility for genetic counsellors to perceive the family *pedigree* as a representation of the disease.

Combination of various types of medical evidence is a powerful fact-mobilising technique, in which clinical workers collectively associate, and thus reconstruct, previously independent, stabilised data. Combinations include large pedigrees resulting from the merging of individual family pedigrees (a process once again leading to the constitution of new scientific-legal objects with an uncertain ownership status), and laboratory pedigrees linking pedigrees to tables of molecular biological data.

Publications are the major modality of *circulation* of combined clinical evidence. They are the result of collective work, one that leads to the establishment, on the basis of an accepted theory, of codified categories of genetic diseases. The role of visual documentation in publications is to produce an integrated image of genetic processes through the combination of relational, external elements with internal images of the patients' bodies. Family histories as reproduced in medical textbooks play an important role in introducing biomedical students to the perception of the family history as relevant to and constitutive of the present expression of illness.

Let us briefly examine in turn each of these elements.

Primary transcription

Primary transcriptions (see Figure 1 for an example) are the first inscriptions derived from oral family stories. Their central characteristic is thus to be a product of interactions between genetic counsellors[3] and their clients. Our analysis of this process will focus on several of its sometimes ambiguous features as embodied in and resulting from the visual representations thus produced.

First, the use of the term 'client' or 'consultant', rather than patient, to designate the persons requesting genetic services is not a slip of the pen, since those persons are not, strictly speaking, patients. The issue of who should be considered a patient (the family, including past generations, an affected child, or the yet to be born children) is fraught with ambiguity. The information elicited by counsellors and represented as a family tree concerns indeed the entire family rather than an individual patient (Richards 1993). Not only is the client requested to provide medical information on other family members, but counsellors can go as far as to suggest that relatives submit to a DNA test. Individual consultations, in return, can affect people other than those who have requested the consultation: 'every member of the family is affected – emotionally, physically, socially – whether patient, at risk, or a spouse' (Gray and Conneally 1993: 90).

If one takes, figuratively, the consultation room as the central terrain on which the ambiguous identity of the client/patient is produced and displayed, then that ambiguity can be said to extend in both directions. On the client/family side, the intricacies of who should count as a family member or, in other terms, the complex interweaving of biological and social issues involved by terms such as family and kinship, have been explored by Richards (1996). Here, let us simply recall that, in Canada, the prototype of the family is the nuclear family consisting of a couple and one or two children (Decima Research 1993), while in Japan the counsellors are confronted with an extended family that includes not only the couple and the children but also the grandparents, who often show up in the consultation room (Ohkura 1989: 9, Kawashima 1992, Saito 1992), and even the ancestors. Furthermore, in Japan, marriage-related issues such as consanguineous marriages are one of the main topics raised during genetic counselling sessions (Ohkura and Kimura 1989, Saito 1992, Fujiki *et al.* 1991). This has an interesting parallel in transcription practices. Among genetic counsellors in Canada, in addition to the 'straight' type of family trees, one also finds a semicircular type, that, according to an interviewee, allows one 'to put as many people on as small a space as possible' (see Figure 2). This argument obviously only applies to the younger generations, since, in contrast, the space available for the older generations is drastically reduced, with the result that semi-circular trees are suitable for constructing families as nuclear rather than as extended social units featuring complex relationships, such as consanguineous marriages.

On the geneticists' side, counsellors are instructed to start drawing a pedigree from the *proband* (or index case), *i.e.* 'the affected individual *through whom a family with genetic disorder is ascertained*', represented as a shadowed icon with an arrow, and then to go back at least two generations in the father's and mother's families (Harper 1993, our emphasis). A clear distinction should thus be made between the proband – the 'first affected family member coming to medical attention' – and the consultand – *i.e.* the 'individual(s) seeking genetic counselling/testing' (Bennett *et al.* 1995). In fact, as shown by observations in the field, genetic counsellors often do not mark the difference between proband and consultand, thus recognising both as being affected by a similar category of illness.

A second feature of primary transcriptions is that they are characterised by the translation of oral family narratives into clinical inscriptions from which all the family data which are not perceived as connected to the specific problem-at-hand are discarded. For instance, one of the procedural guidelines for taking pedigrees in a general hospital we surveyed suggested that genetic counsellors 'may use a diamond to summarize, if family members do not have noteworthy problems and are greater than first degree relatives'. Counsellors are also warned that 'in obtaining the family history one should ask about the same or related diseases found in the index case or patient, rather than about a list of diseases of "familial tendencies"' (Gelehrter and Collins 1990: 259). Interestingly, this feature is linked to another ambiguity characterising the production of primary transcriptions,

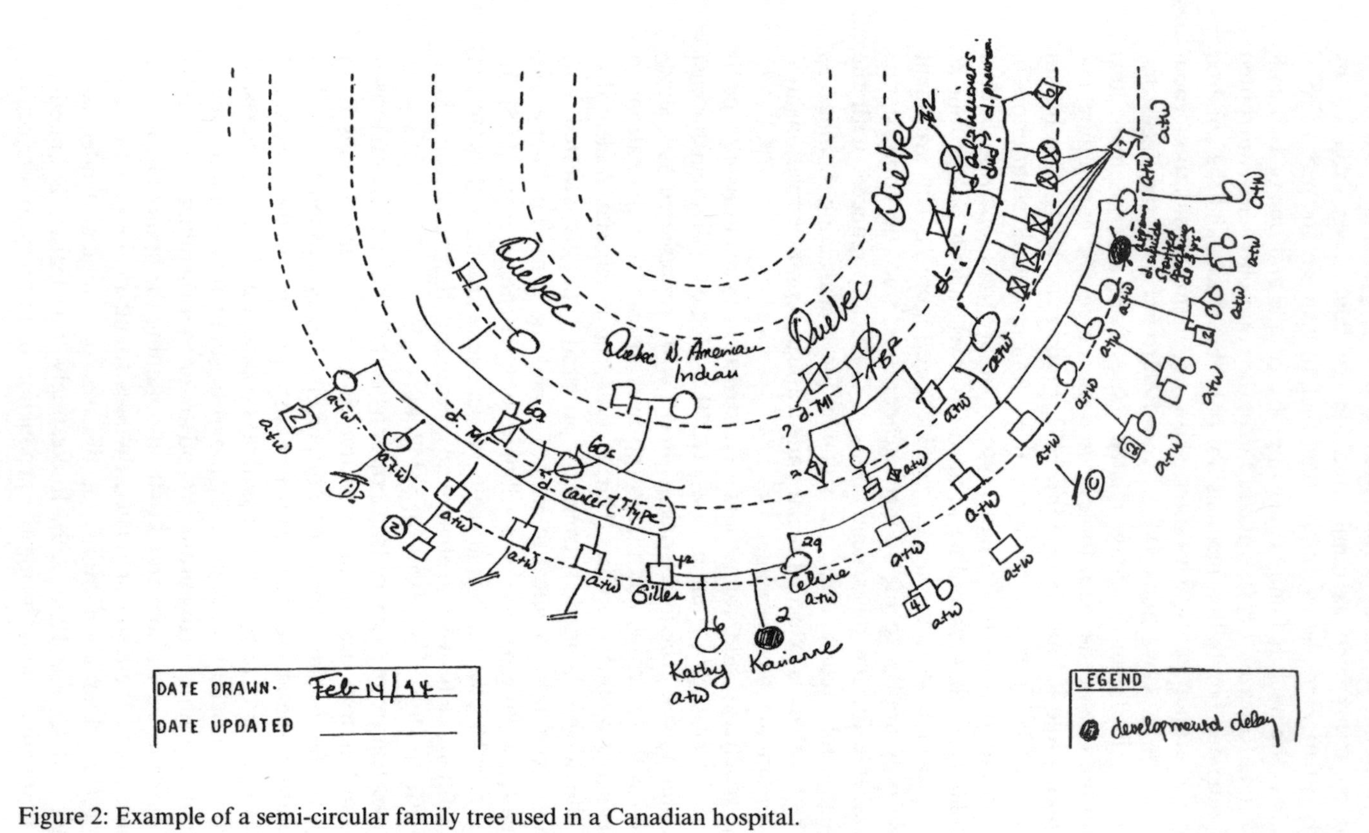

Figure 2: Example of a semi-circular family tree used in a Canadian hospital.

i.e. the fuzzy nature – sociological or biological – of the data thus represented. For sure, in drawing a family tree and listing the client's near relatives by sex, age, state of health, and, in Canada, religion and ethnic background, particular attention is devoted to the occurrence of relevant illnesses in the family. The resulting products can therefore be said to correspond to a 'biological' family history (Bennett *et al.* 1995), but insofar as they amount to a collective representation of the family illness or, in other words, of the family as illness, they undeniably display a social dimension.

Questionnaires used during the intake process in Canadian hospitals contain two types of questions: demographic questions (*e.g.* place of living, type of work, number and age of children and relatives), and questions about health and illness (*e.g.* reasons of death of parents and relatives, presence of certain diseases). It is thus legitimate to ask whether family trees could not also be conceived of as sociological tools and, indeed, both Western and Japanese genetic counsellors have explicitly pointed to the simultaneously sociological and biological nature of family trees. Bennett and Steinhaus, for instance, noted that 'the family history is the most powerful tool of the genetic counsellor. It serves not only as a diagnostic tool, but also as a *sociological* aid in counselling by serving as a record of family relationships' (1992: 312, our emphasis), and Fujiki characterised one of his major research projects as involving the investigation of the 'biological and sociological characteristics of an isolated community' (1981: 4).[4]

A third feature of primary transcriptions, one that brings us back to the previously evoked tension between the individual clients and their family, concerns the map-like properties of family trees. Given the widespread occurrence of the 'map' metaphor in human genetics, as embodied in expressions such as 'mapping the human genome' and in claims such as Jones's that 'genetics, like geography, is about maps' (1993: 41), we feel warranted in exploring this analogy. On a family tree, as well as on the medical pedigrees derived from them by secondary transcriptions, an individual body is represented by a symbol such as a square (for males) or a circle (for females), black in the case of an affected individual, white in the opposite case. Different varieties of shading and, sometimes, colour, provide additional symbolic tools further to differentiate and contrast individual bodies/icons. As in a map, each individual symbol, taken separately, has a low information content (Nakayama 1990), the meaning residing in the associative network of icons. Thus, as is the case with maps, family trees not only represent selected aspects of reality, but do so in a way that produces a new representation of the family and of the individual's relation to it. As such, they operate along the dual lines described by Wood, *i.e.* both 'as a medium of language [. . .] a visual analogue of phenomena' and 'as a myth [that] refers to itself and its makers' (1992: 116). The visual connections between icons promote the idea of the family as a whole (as noted by a genetic counsellor: 'everything about the family is there on paper') and of the illness as a property of the whole family network. In this sense, family trees refer to a collective, diseased body and they become a visual memory of the family disease.

It is, however, not only the representational dimension of the map analogy that is of interest to us here, but also its performative dimension. As noted, among others, by Olson (1996), maps replace deictic definitions of one's location ('I am here') with a 'single picture of the world', thereby providing a new, coherent point of view from which the world can be explored. A map acts as 'the model or theory of which the voyages [of discovery] are the empirical test' (1996: 211–12). Similarly, in the search for genetic explanations of hereditary diseases family trees act as maps in the location and interpretation of DNA samples from affected individuals. Family trees externalise not only visible disease manifestations, but also the invisible, not manifest illness of a 'carrier' as identified by a partly shaded icon. Thus, they may even lead to the 'overproduction' of family diseases, by promoting, for instance, the existence of new pathological categories such as 'disease carrier' (Markel 1992).

A final feature of primary transcriptions concerns the extent to which they bear the mark of the local setting within which they were produced or the local practitioner who produced them. The skilful, artisanal qualities of primary transcriptions are reflected in the common wisdom among practitioners that the production of a family tree is part of the 'art' of a given counsellor. To be sure, local variations can often be linked to institutional factors, such as the locus of training. For instance, Japanese genetic counsellors utilise different types of symbols according to whether they had been trained as human geneticists or medical geneticists. The former follow the style and methods of their Western colleagues as reported in textbooks such as Thompson *et al.* (1991) and Harper (1993). The latter resort to the symbols and approach developed by Dr. Ohkura, the leader of the medical geneticists' association (*e.g.* Uchida 1987). Thus, in the Japanese case, to resort to a given set of symbols is to mark one's professional identity and allegiance to a particular school of thought.

In spite of these regularities, however, a large variety of formats is still found to exist in Canada and Japan. For instance, in North America pregnancy is symbolised in at least 17 different ways, with different meanings being attributed to some of the same symbols (Bennett *et al.* 1993). A Canadian MD noted that 'as long as you put a mark in the legend box, you can use any kind of symbol you want'. A genetic counsellor claimed that whenever she looked at a pedigree, she could easily imagine who, in her department, had done it. The diversity of formats and symbols used in primary transcriptions are a potential threat to the performance and expansion of genetic counselling. Medical activities are made possible by various forms of regulation, including standardisation (Keating and Cambrosio 1997). Genetic counselling is no exception: secondary transcriptions are part of the answer to the problem raised by the local variability of primary transcriptions.

Secondary transcriptions

As opposed to primary transcriptions, which, as we have seen, result from interactions with family members, secondary transcriptions (*i.e.* edited versions of family trees; see Figure 3) are mainly the outcome of interactions between genetic counsellors and other associated health care workers, and are thus to a large extent regulated by clinical norms and guidelines.

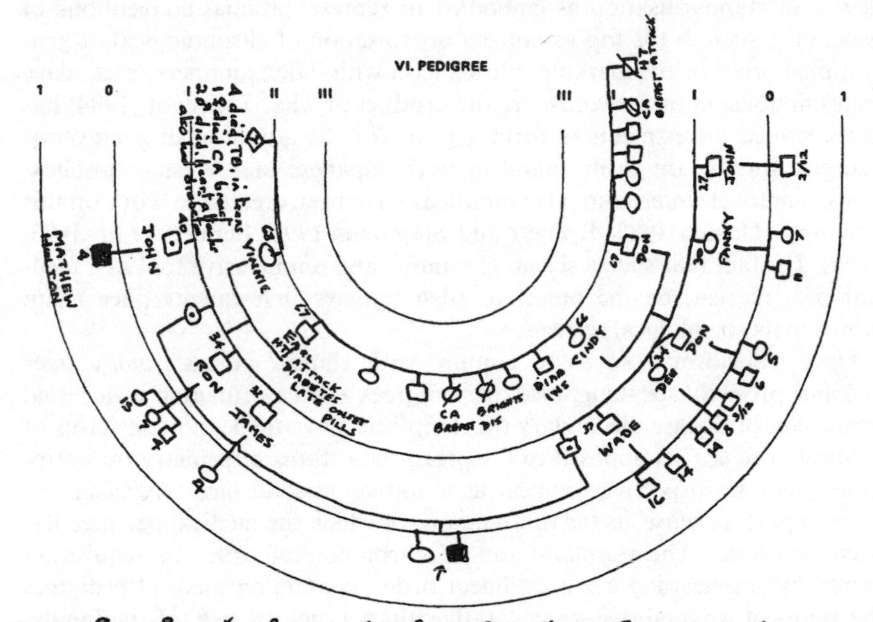

Record — abortions, stillbirths, cancer, cong. defects, diabetes, epilepsy, short stature, twins, mental retardation, conditions like probend's

Pedigree Number	Name	Address, hospital, diagnosis, confirmation of diagnosis, etc.
0-1	MATHEW HOLTON	CYSTIC FIBROSIS
1-2	VINNIE (ROSEMARY)	DIABETIC. DX 40's. INSULIN NOW FOR 10 YEARS. HAD
		HYPERACTIVE THYROID. RX NOW ON PILLS
1-10	CINDY WING	PGM HAS HEART PROBLEM. ON LOTS OF DRUGS. HAD
		ANGINA → HEART ATTACKS X2

Figure 3: Example of a secondary transcription used in a Canadian hospital; compare with Figure 1.

The purpose of secondary transcriptions is not further communication with the family, but, rather, the sharing of information with other practitioners. Indeed, not only is 'the construction of an accurate family pedigree [. . .] fundamental to the provision of clinical services'; it also 'serves as an informational framework for human genetic research' (Bennett *et al.* 1995: 745). Secondary transcriptions are therefore expected to exhibit a certain degree of standardisation as embodied in representational conventions of generality, such as the top-to-bottom organisation of chronological or generational order or the marking of each icon with index numbers. Secondary transcriptions, in other words, are the product of what Thévenot (1984) has aptly termed 'investments in forms'. Calls for the adoption of a universal pedigree format are easily found in both Japanese and Western publications; moreover, international committees have been created to work on this issue (*e.g.* Uchida 1987, Bennett and Steinhaus 1992, Bennett *et al.* 1993, 1995). The fact that such a seemingly simple and unobtrusive tool as a medical pedigree can be the object of such massive investments once again points to its sociological relevance.

Once transformed so as to comply with clinical norms, *family* trees become, properly speaking, *medical* pedigrees and can function as a visual expression of disease. Secondary transcriptions have the legitimate status of 'clinical evidence' as opposed to the proto-data status of primary transcriptions which retrospectively appear as tentative, unconfirmed renderings or, in an important sense, as the *raw data* out of which the medical pedigree has been produced. The normalisation of chronological order in requisition forms, by emphasising a visible, linear order, confers on medical pedigrees the status of a visual case *record*, rather than a mere *sketch*, of the family, for which genetic counsellors are usually required to assume accountability by signing. The term 'record' should not be understood, in this respect, as the passive outcome of recording activities. Rather, as noted by Raffel (1979) 'when records are seen in terms of the grounds which make them possible, it is no longer adequate to state that records reflect, whether accurately or inaccurately, the givens of the real world, because the real world itself comes to be shaped by the very idea of recording it' (1979: 48–9).

By reordering and stabilising the initial family data, genetic counsellors shift their focus from individual family members to the medical pedigree. Indeed, an often cited aspect of the medical pedigree is its 'at a glance' quality. As noted by a Japanese public health nurse, 'the most important thing is to select the information in order for the other associates to see and understand it at a glance'. Secondary transcriptions increase the distance between the abstract icons and symbols used in family trees and the concrete contexts of primary data collection, impose an abstract chronological order and end up producing an objective image of family diseases. By attaining the anonymity of clinical evidence, this image becomes 'generally available to all who share, or may share in the future, in the sign system in question' (Berger and Luckmann 1966: 64). In fact, medical pedigrees act as intermediaries between, and thus as objects for co-ordinating the action of, various kinds of health-care professionals within the genetic counselling team or centre. They also act as intermediaries between the latter and external practitioners.

Genetic counselling is not always carried out by a single practitioner, but often, especially in Canada, by a teem of several counsellors, who operate sequentially to check and complete the pedigree. The team may include not only genetic counsellors, but also, for instance, biologists whose co-operation in the collective endeavour is mediated by the pedigree. During case conferences in a Canadian children's hospital, genetic counsellors draw the pedigree on the whiteboard, other participants then modify the pedigree drawn by the speaker and the pedigree thus becomes the actual focus of discussions centred on deciding whether 'the family history is positive or negative' in terms of diagnosis. Pedigrees, in these situations, fit Suchman's (1990: 314) insightful discussion of diagrammatic, shared conceptual tools around which situated interactions develop.

Pedigrees play a similar co-ordinating role between genetic counsellors and other associated workers in different departments or hospitals, such as specialists in rare genetic diseases or supervisors of diagnostic laboratories. As noted by Gelehrter, 'the history must be recorded so that it communicates information to all health professionals caring for the patient' (1983: 122). This requirement has been embodied in material devices such as referral forms. Thus, molecular diagnostic laboratories require genetic counsellors to complete a requisition form containing a pedigree that must provide details supporting the reason for referral, be it a documented family history of a given disease, a possible family history of a given disease or symptoms of a given disease in an individual. In other words, the pedigree on the requisition form does not simply show family relationships, but, more importantly, points to the presence of the indicated disease. Thus, to return to our previous map analogy, the pedigree, prior to DNA testing, highlights the relationship between the disease and the family and points to the elements to be explored by providing signposts such as the relationship with the index case, a clinical construct produced by the transcription process.

In addition to promoting family trees to the status of clinical evidence, secondary transcriptions provide medical pedigrees with a legal-administrative status that makes them the object of further investments and allows them in turn to function as a tool for regulating practices. The Canadian College of Medical Genetics (CCMG), for instance, has established a system of guidelines and accreditation for genetic centres that requires them to maintain clinical records, including a 'basic, but conventional pedigree [that should] at least [extend] to second degree relatives' (unpublished CCMG document). Pedigrees have become part of the larger network of clinical case records and have to adapt to their format. For instance, hospital files of a Montreal hospital require the use of semi-circular medical pedigrees and genetic counsellors who prefer to produce straight family trees have subsequently to transform them into a semi-circular format.

Routine genetic counselling practices, such as the construction of family trees and medical pedigrees, produce visual representations of family data that operationally redefine the notions of family, population and disease. Indeed, genetic counselling practices do not simply involve a process of simplification of culturally complex family data, they also mobilise the construction of new evidence that, in turn, creates new entities and new discursive domains, such as the debates surrounding the ownership of 'family data' (Frankel and Teich 1993). This is clearly shown by the analysis of combination processes.

Combination

Combination is the process of articulating, or 'triangulating' (Star 1986), a medical pedigree with other evidence, such as other pedigrees or results of genetic tests. The resulting, composite image acts as robust evidence for the presence of a 'genetic disease' and, as such, counteracts the openly acknowledged uncertainty (*e.g.* Fraser 1988: 204) surrounding medical diagnoses in genetic counselling. The process of combination is a powerful technique for mobilising previously stabilised facts (Latour 1990). Whereas primary and secondary transcriptions are characterised by the genetic counsellors' efforts to translate and stabilise complex family data in a clinical context, combination is defined by the collective work of mobilising and managing facts across different clinical settings. As such, combinations function as 'boundary objects' (Star and Griesemer 1989) that allow for divergent uses, interpretations, and reconstructions to be practised by the various health-care workers.

By combining individual pedigrees obtained from different sources, a so-called 'large pedigree' is produced that is most often used as part of a research project or a screening programme for 'family diseases' such as Huntington's Disease and colon cancer, and is kept as a new record, separated from individual case records. For instance, in a general hospital in Canada, following the development in the late 1980s of a screening programme for Huntington's Disease, clinical charts and databases have been stored in the local computer network, leading to the rapid accumulation of data including many individual family pedigree files which have been combined into larger pedigrees.

Because of confidentiality issues, individual patients and family members are often not aware of the content of large family pedigrees. A member of the Huntington's screening programme noted that:

> What we do in Huntington's cases is surely tricky. Usually, when I was speaking to the family with Huntington's Disease, I would certainly show

> them a pedigree, but that turned out not to be a good thing to do, because everybody understands the things we add to a pedigree. But it turns out that maybe a family member A did not know what family member B had told us, and so he will be suspicious of how we got the information. What we do now is to construct a pedigree for each person who comes in, and that information is taken from their point of view. And then we put together a general pedigree and we don't show it to anybody else in the family. Each one has a separate chart, and we may have a big, big chart.

The quotation points to the novelty of 'family data', that appear as both the outcome of pedigrees and a new form of reality, with wide-ranging legal, social and biological consequences. This is even more obvious when one considers the question of the ownership of family data. As noted by Cook-Deegan: 'The questions [of how to make the pedigree public] are far from clear, . . . since information is neither fully private nor fully public, but often somewhere in between' (1994: 85). The ambiguity in regard to the public/private dichotomy is grounded in the uncertain status of 'family information', or, more precisely, in the way the latter is produced. As previously discussed, the information derives from and refers to more than a single person, thus challenging simple solutions such as the signing of informed consent forms by individual patients (Powers 1993, Frankel and Teich 1993). Moreover, a large pedigree resulting from a research project is a collective production of a set of persons including various family members, genetic counsellors, geneticists, and laboratory workers, or even associations of affected individuals and public foundations. It is thus impossible for the final result to be attributable to a single individual or group of individuals. The point, here, is that large pedigrees lead to open-ended extensions of established social, legal and biological configurations and, in so doing, participate in the redefinition of some of the distinctions (*e.g.* body/society, normal/pathological) in which our understanding of social life is grounded.

The accumulation of clinical data produces not only intentional research combinations, but also coincidental combinations, thus again pointing to the fact that pedigrees, rather than a mere record of known facts, are non-trivial objects, leading to the production of a new reality. For instance, a genetic counsellor in a children's hospital told the story of a woman suffering from Myotonic Dystrophy whose affected relative had not been tested in the lab. When her blood sample was sent to the DNA lab for genetic testing, the lab called back to report that their files already contained data stored under the same family name. Were they related? The counsellors and the lab practitioners each pulled out their pedigree, which turned out not to include birth dates, uncles and cousins because they had not been deemed relevant to the counselling situation. The patient was called back, the relevant details filled in and the updated version of the pedigree was then compared with the pedigree in the laboratory. In this particular case, they did not match at all, but had they matched to some extent, then further decisions would have had to be taken on how many names and dates of birth would be needed before an actual match could be declared.

The accumulation of individual pedigrees enhances the opportunities for associated health-care workers to communicate with each other, to 'discover' resemblance between family pedigrees, and to engage in the construction of new, large pedigrees. In turn, the availability of large pedigrees has feed-back effects on the initial process of family data collection, thus showing the ongoing, recurrent extension of this tool. For instance, with the unfolding of screening programmes, and in spite of the difficulties involved in recording detailed lists of names and dates of birth, genetic counsellors now insist on the importance of registering this kind of information and new local policies to this effect have been drafted by research teams and hospital authorities.

The production of a large, combined pedigree is a difficult and time-consuming endeavour that often involves the painstaking accumulation of information elicited from individual genetic counselling sessions, but that can also resort to available medical and administrative records. A condition of possibility for this latter approach is, of course, the presence of a relatively stable population over long-term periods, a condition better realised in Japan than in Canada (but see, for the case of Quebec, Bouchard and De Braekeleer 1991). Thus, the Huntington's Disease screening programme initiated by a general hospital in Canada collected family data mainly by genetic counsellors' intake, whereas Kanazawa (1994) reported that his research group investigating Huntington's Disease in Japan was able to establish a data bank of about forty family pedigrees by approaching many practitioners. Ethical requirements restricting access to administrative files for genetic research purposes are another important factor, one that, again, seems to favour Japanese geneticists who can easily access so-called household registers (*koseki*) maintained by local administrations.[5]

An interesting example of the construction of a large pedigree in Japan is offered by Fujiki's (1981) work on an isolated village with a high prevalence of Laurence-Moon-Biedl syndrome, a rare autosomal recessive disease. Data collection mainly resorted to a variety of official records, ranging from present-day resident cards to the death registers maintained by Buddhist temples. By various graphical manipulations such as establishing vertical connections between pedigree sheets with the same surname and horizontal connections between family units with the same parents, Fujiki (1981: 3) managed to ascertain 'all blood relationships between individuals, including those deceased, [. . .] resulting in a population three or four times as numerous as the present population'. The final diagram, reconceptualised as a 'village pedigree' (see Figure 4), visualises a notion of population as defined by common ancestors obtained through the calculation of the degree of inbreeding. A medical, and yet, simultaneously, a social and mythical object, – to which its aesthetic qualities (a finely-carved, expanding circle) are possibly not foreign – the pedigree has become a new symbol for the community: a representative of the village asked the geneticist for a copy of the pedigree and the village was formally presented with it as a gift (interview with Dr. Fujiki).

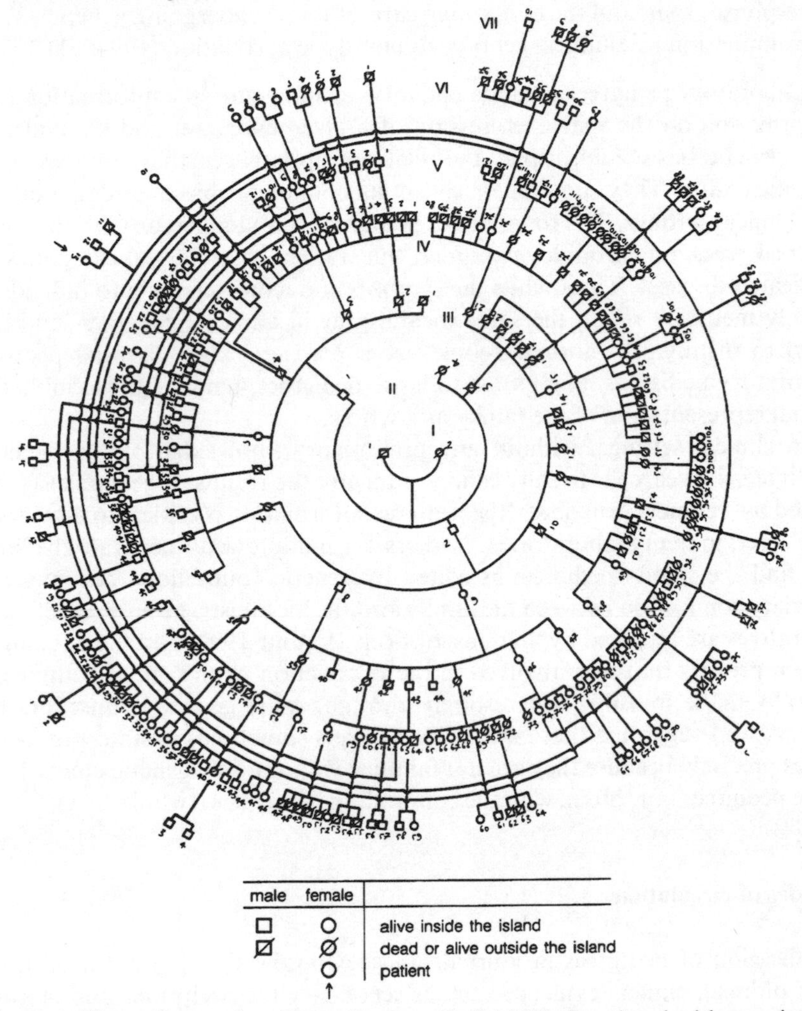

Figure 4: The pedigree of a village. (Source: Fujiki 1981. Reprinted with permission from Dr. Norio Fujiki.)

In contrast to large pedigrees that combine the same type of evidence, 'lab pedigrees' (a colloquial term used by genetic counsellors and molecular biologists) involve the combination of pedigrees with tabulated results of genetic or DNA tests. As clearly stated by Billings, the triangulation of these different types of evidence lies at the core of modern human genetics:

> human genetic information is more accurately defined as the composite product of several data sources and processes including: the elucidation of family history; the examination of genetically related phenotypic manifestation in family members; the results of biochemical, cytogenetic and DNA analytic testing; the genetic counselling of families and their responses to it; and the continuing care of those undergoing genetic examination to establish genotype/phenotype correlation (1994: 51).

Laboratory pedigrees conflate phenotypic and genotypic information (*i.e.* information on the visible expression of a given character and its 'underlying' genetic basis) and display the relationship between these two levels of genetic reality. They simultaneously act as boundary objects bridging different clinical settings. The focus of lab pedigrees is no longer, as with the clinical pedigrees, on chronological order, but, rather, on indexical categories of indicated diseases. Thus, while the symbols and icons referring to individual family members retain the same meaning as in clinical pedigrees, and are used to distinguish 'normal people' from 'carriers' and 'diseased people', laboratory pedigrees, in contrast to large pedigrees, tend to disassemble the visual representation of the family as a whole.

In clinical settings, without any preliminary knowledge of the medical pedigree, a researcher usually cannot interpret the results of genetic tests. As noted by a clinical geneticist: 'the pedigree information is critical to any DNA test'. Yet, in many laboratories, workers are not allowed to contact the family, and the family pedigree, as edited by genetic counsellors, becomes the primary connection between the family and the biologists. Family *and* clinical narratives are replaced by 'flat' inscriptions (Latour 1990) and the documentation process that was involved in the production of the initial pedigree is thereby made invisible. Interestingly enough, some genetic counsellors we interviewed suggested that laboratory workers sometimes misinterpret pedigrees precisely because they are not familiar with the way genetic counsellors take pedigrees nor, often, with the counselling process as a whole.

Modes of circulation

Publication of pedigrees in journals and textbooks signals the transformation of local, clinical evidence such as secondary transcriptions and laboratory pedigrees, into certified biomedical evidence circulating in extended networks. As expected, this step further down the inscription cascade is marked by new processes of standardisation. It also leads to renewed negotiations among genetic counsellors, pedigree researchers, clinical laboratory workers, journal editors and even family members around such topics as the confidentiality and ownership of the published information, that, once again, point to the open-ended nature of pedigrees.

As is the case with other forms of visual representations, medical journals, such as the *American Journal of Medical Genetics*, require prospective authors to follow a fixed format for publication and to use specified pedigree symbols. Computer software used in the production of manuscripts also increasingly contributes to this standardisation process. A quick look at journals, however, suffices to show that the goal of 'universal standardisation' called for by various committees from early on in this century (*e.g.* Carr-Saunders *et al.* 1912–13) to recent times (*e.g.* Steinhaus *et al.* 1995), is far from having been achieved. Publication conventions often differ from the ones used in local settings, and these differences can be accounted for by the different use to which the pedigree imagery is put: the greater the distance from actual clinical interventions, the greater the degree of abstraction of pedigrees: a person's actual age, for instance, will be replaced by a number referring to 'the person's number within that generation' (Gelehrter 1983: 122).

One aspect of this process of increasing abstraction is of particular interest since it involves an issue often referred to by pedigree researchers as the necessary trade-off between confidentiality and the accuracy of scientific data. Pedigrees can be altered to secure anonymity but, because of the breadth and depth of information conveyed by pedigrees, it may be difficult to preserve anonymity without jeopardising accuracy. Some counsellors have developed a few 'tricks' to modify published family pedigrees. Frankel and Teich, for instance, point to the long-standing practice in medical case reports of 'disguising the identities of study participants by, for example, changing gender or age, or shifting birth order' (1993: 31–2) as equally applicable to pedigree studies. Simpson, however, argues that the practice of disguising the sex of family members 'may decrease the scientific value of the pedigree' (1993: 49).

But what, exactly, is the 'scientific value' of pedigrees? How, in other words, do they contribute to scientific and clinical arguments? Publications featuring pedigrees more often than not combine them with other kinds of genetic imagery, thus presenting an integrated, coherent picture of genetic disease. For example, the integrated picture of a 'family disease', shown in Figure 5, used during an intensive genetic counselling training seminar, was published in the Japanese journal *Knowledge of Medicine* under the title 'Seeing Genetic Disease: Fragile X Disease'. The photographs represent both the inside and the outside of the body. The external body photographs show a front and side view of the external facial appearance and a front view of the genitalia. The internal photographs consist of three chromosome pictures. The first one represents the first proband case in the upper family history. The second one shows the proband in the second family history. The third one is that of the carrier. In addition to these body photographs, the figure shows the diagrams of two 'family histories' whereby the presence of the illness is signalled by a different shading of the icons. These different types of visual representations, taken separately, account for some aspects of this particular manifestation of the Fragile X disease. It is only, however, by juxtaposing the various elements separately produced by collaborating researchers, that the illustration constructs an objectified, aggregated image of the disease.

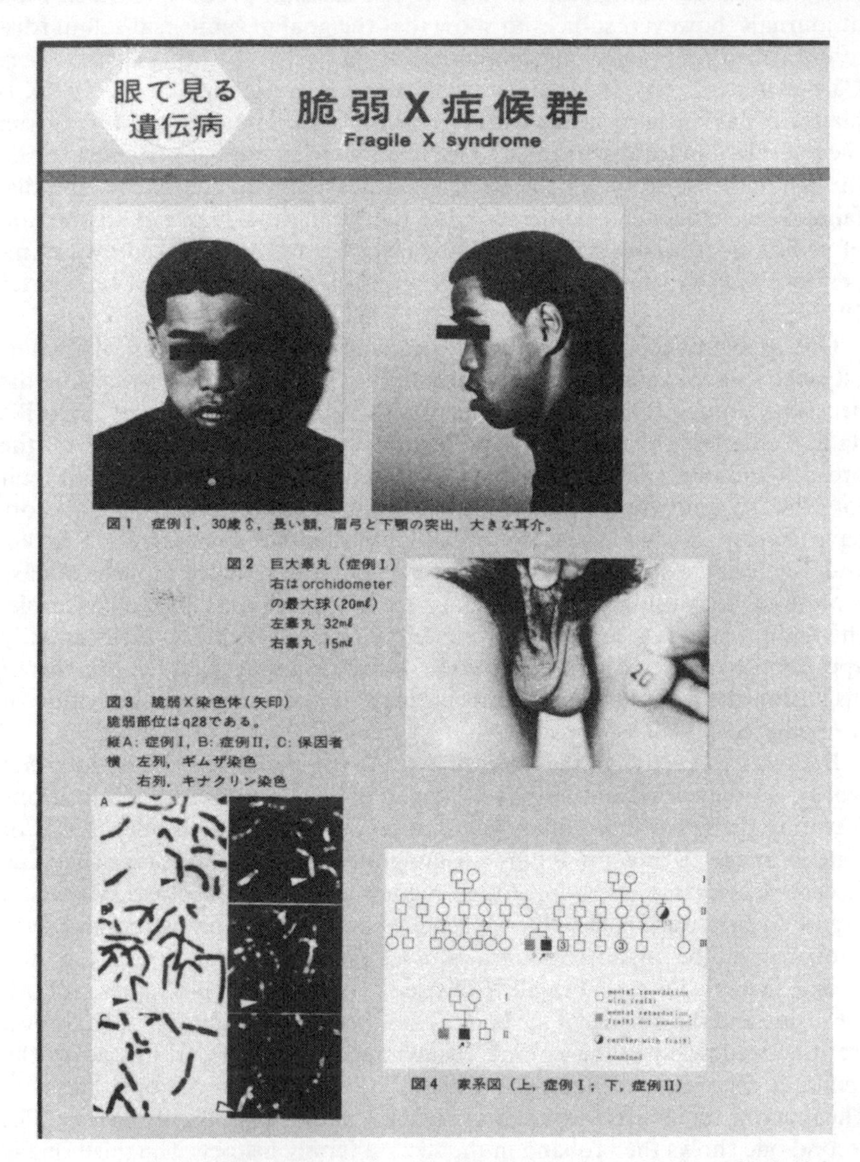

Figure 5: An integrated picture of a 'family disease' (Fragile X) published by a Japanese medical journal. (Source: Ishikiri and Niikawa 1984. Reprinted with permission from Life Science Publishing Co., Ltd.)

What is the specific role of the pedigree in this respect? We would suggest that pedigree imagery, rather than simply being one of the visual elements that add up to the composite image of the disease, plays a strategic mediating role. The variously shaded icons, because they are icons of a collective body, work as an integrative tool associating representations of the internal, allegedly causal features of the genetic disease with representations of its external manifestations. In the process of producing these associations, the family history and the external body features are earlier than the internal images; however, the juxtaposition of these pictures presents all these elements as simultaneous. In turn, through the juxtaposition and triangulation of various forms of medical imagery, the family history is recognised as part of the present disease.

This process culminates in medical textbooks, which often resort to medical pedigrees to illustrate the various (single-gene, multiple-gene) transmission patterns of genetic diseases. In doing so, they use highly schematic pedigrees, from which all elements that would allow for their use in concrete clinical situations have been erased (*e.g.* Harper 1993: 20). This procedure leads to the production of what Fleck calls 'ideograms', *i.e.* 'graphic representations of certain ideas and certain meanings', whereby 'the meaning is represented as a property of the object illustrated' (1979: 137). Textbook pedigrees become, in fact, the embodiment of genetic disease – they stand for the latter – and play an important role in the socialisation of geneticists and genetic counsellors. Before clinical workers can even start to treat family members in the counselling sessions, they have internalised a representation of genetic diseases that takes the form of a medical pedigree.

Conclusion

The focus of this chapter has been on medical pedigrees as mediating devices between external and internal, collective and individual aspects of the body, as well as between different professional practices. We have examined the sequential transformation of family trees into medical pedigrees, *i.e.* into tools for the production of medical evidence, or, more precisely, for the performance of family diseases and, further on, as icons of 'genetic disease'. Medical pedigrees should not be reified: their power as clinical tools lies less inside them than in the fact that they can be mobilised as part of an expanding network of molecular evidence. Yet, one should also be careful not to 'oversocialise' pedigrees: objects and tools play a mediating role in social relations, to which they cannot be reduced (Berg 1996, Latour 1994).

Indeed, one of the implications of our analysis is that while pedigrees cannot be treated as an objective rendering of a reality, they cannot either be simply conceived of as the result of a 'social construction' if, by the latter, one claims that their constitution and meaning are entirely accountable in terms of social factors. Just as pre-existing notions of kinship relations inform pedigrees, so, too, do pedigrees contribute to the constitution and circulation of new understandings and renderings of kinship as embodied in notions such as family disease and issues such as the ownership of family information. There is no split, here, between objects and subjects, but, rather, from the very beginning, an intimate, mutually-constitutive relationship.

A second implication of our analysis points to the issue of the novelty of the 'new genetics'. While we do agree with several aspects of the argument according to which the application of molecular biology techniques to medicine constitute a radical change of perspective (Rheinberger 1995), we hope to have shown that the new genetics and its clinical performance resort to apparently 'mundane' tools such as pedigrees, that have a longer, more complex history. This also means that the 'new genetics' is not exclusively grounded in laboratory tools and materials, but also resorts to clinical data and records: laboratory samples will not work properly without the accompanying pedigree. For sure, the new genetic network into which pedigrees are by now integrated is a different network from the ones of which they had previously been part. Yet, these differences should not blind us to the presence of continuities inscribed in the very tools that are transmitted from one period to the other.

Finally, and in relation to our previous point, we hope to have shown that a central issue in relation to which these continuities can be explored is the notion of the family. Quite often, medical sociologists and historians have approached the topic of human and medical genetics in terms of the twin issues of race and eugenics. The association between pedigrees and family, while certainly at first less dramatic, appears to us as at least as decisive for any detailed exploration of the contemporary mechanisms of 'bio-power' and 'bio-sociality'.

Acknowledgements

This article is based on the first author's M.A. Thesis in Medical Sociology (The Visual Transcription of 'Family Disease': A Comparison of the Use of Medical Pedigrees in Genetic Counseling Practices in Canada and Japan, McGill University, 1995), written under the supervision of the second author. The thesis has greatly benefited from Margaret Lock's anthropological insights into Japanese health care practices and Abby Lippman's comments on several aspects of genetic counselling. Peter Keating and Jean-Paul Gaudillière read two subsequent versions of this paper and offered thoughtful suggestions and corrections. We are, however, solely responsible for the opinions and any errors and inaccuracies contained in this chapter. We should like to thank the medical geneticists and genetic counsellors who agreed to be interviewed and observed, gave us access to personal files and documents, and commented on preliminary versions of the chapter. Alberto Cambrosio's ongoing research on medical technologies is supported by Social Sciences and Humanities Research Council of Canada Grant 410-94-0352 and Fonds FCAR Grant 95-ER-2220.

Notes

1 In this and following figures, names and other details have been changed so as to preserve anonymity. See our discussion of this issue in the section 'Modes of circulation'.
2 All observations were carried out by the first author, who was granted access to training sessions, counselling sessions and team conferences.
3 Genetic counsellors include, in Japan, MD counsellors and public health nurses, and, in Canada, Ph.D. or MD geneticists as well as so-called genetic associates, *i.e.* health-care workers who have either a background in counselling, such as social work or psychology, and learn genetics on the job, or those who have training in nursing, genetics, or other paramedical skills. In both cases, and in spite of exceptions due to personnel shortages, there is a division of labour, with, in Japan, public health nurses and, in Canada, genetic associates, specialising in the taking of family trees, while, in both cases, MD geneticists specialise in the diagnosis. One should note here the parallel development of the pedigree as a tool, of new forms of medical intervention and of the division of medical work.
4 The ambiguous status of family trees – biological or sociological? – goes hand in hand with the ambiguities surrounding the definition of genetic counselling devised in the 1970s by a committee of the American Society of Human Genetics and stating that: 'Genetic counseling is *a communication process* which deals with the human problems associated with the occurrence, or the risk of occurrence, of a genetic disorder in a family' (Fraser 1974: 637). The definition points to an essential tension in the historical development of this practice. The first generation of genetic counsellors conceived of their expertise largely as one of presenting medical and scientific facts. Since the mid-1970s, however, many articles have criticised the medico-scientific orientation of genetic counselling and the lack of a psycho-social perspective (Terrenoire 1986). Whether the definition could be seen as the triumph of 'the psychological paradigm shift in genetic counselling' (Kessler 1980), or as the embodiment of a division of labour in genetic counselling, whereby technical knowledge still plays a primary role and the psychological perspective is entrusted to lower-status members of the clinical group (Yoxen 1982), remains an open question.
5 As suggested by a Japanese population geneticist, while this was indeed the case until about 10 years ago, it has now become more difficult to carry out this kind of research.

References

Amann, K. and Knorr-Cetina, K. (1990) The fixation of (visual) evidence. In Lynch, M. and Woolgar, S. (eds) *Representation in Scientific Practice*. Cambridge, MA: MIT Press.

Atkinson, P. (1995) *Medical Talk and Medical Work. The Liturgy of the Clinic*. London: Sage.

Bennett, R.L. and Steinhaus, K.A. (1992) Standardization of the family pedigree, *Journal of Genetic Counseling*, 1, 312–13.

Bennett, R.L., Steinhaus, K.A., Uhrich, S.B. and O'Sullivan, C.K. (1993) The need for developing standardized family pedigree Nomenclature, *Journal of Genetic Counseling*, 2, 261–73.

Bennett, R.L., Steinhaus, K.A., Uhrich, S.B., O'Sullivan, C.K., Resta, R.G., Lochner-Doyle, D.L., Marker, D.S., Vincent, V. and Hamanishi, J. (1995) Recommendation for standardized human pedigree nomenclature, *American Journal of Human Genetics*, 56, 745–52.

Berg, M. (1996) Practices of reading and writing: the constitutive role of the patient record in medical work, *Sociology of Health and Illness*, 18, 499–524.

Berg, M. and Mol, A. (eds) (1997) *Differences in Medicine. Unravelling Practices, Techniques and Bodies*. Durham, NC: Duke University Press.

Berger, P.L. and Luckmann, T. (1966) *The Social Construction of Reality: a Treatise in the Sociology of Knowledge*. New York: Doubleday and Company.

Billings, P.R. (1994) Genetic information in the health care reform era. In Macer, D.R.J. (ed) *Bioethics: for the People by the People*. Tsukuba: Eubios Ethics Institute.

Bosk, C.L. (1992) *All God's Mistakes: Genetic Counseling in a Pediatric Hospital*. Chicago: The University of Chicago Press.

Bosk, C.L. (1993) The workplace ideology of genetic counselors. In Bartels, D.M., LeRoy, B.S. and Caplan, A.L. (eds) *Prescribing our Future: Ethical Challenges in Genetic Counseling*. New York: Aldine De Gruyter.

Bouchard, G. and De Braekeleer, M. (eds) (1991) *Histoire d'un génome. Population et génétique dans l'Est du Québec*. Sillery: Presses de l'Université du Québec.

Carr-Saunders, A.M., Greenwood, M., Lidbetter, E.J. Schuster, E.H. and Tredgold, A.A. (1912–13) The standardization of pedigrees: a recommendation, *Eugenic Review*, 5, 66–7.

Cook-Deegan, R.M. (1994) Ethical issues arising in the search for neurological disease genes. In Fujiki, N. and Macer, D.J.R. (eds) *Intractable Neurological Disorders, Human Genome Research and Society. Proceedings of the Third International Bioethics Seminar in Fukui, Japan, 19–21 November, 1993*. Tsukuba: Eubios Ethics Institute.

Decima Research (1993) Social values and attitudes of Canadians toward new reproductive technologies. In *Royal Commission on New Reproductive Technologies, Research Volume 2*. Ottawa: Minister of Supply and Services Canada.

Duster, T. (1990) *Backdoor to Eugenics*. New York: Routledge.

Fleck, L. (1979) *Genesis and Development of a Scientific Fact*. Chicago: University of Chicago Press.

Foucault, M. (1977) *Discipline and Punish. The Birth of the Prison*. New York: Pantheon Books.

Frankel, M.S. and Teich, A.H. (eds) (1993) *Ethical and Legal Issues in Pedigree Research: Report on a Conference Sponsored by the AAAS Committee on Scientific Freedom and Responsibility and the AAAS-ABA National Conference of Lawyers and Scientists*. Washington, DC: American Association for the Advancement of Science.

Fraser, F.C. (1974) Genetic counseling, *American Journal of Human Genetics*, 26, 636–69.
Fraser, F.C. (1988) Genetic counseling: using the information wisely, *Hospital Practice*, 23, 245–66.
Fujiki, N. (1981) Using family linkages to reconstruct an isolated Japanese villager's history. *World Conference on Records, Vol. 11. Asian and African Family and Local Family History*. Genealogical Society of Utah.
Fujiki, N., Kishi, K. and Hirayama, M. (1991) Japanese perspectives on ethics in medical genetics. In Fujiki, N., Golyzhenkov, V. and Gankowski, Z. (eds) *Medical Genetics and Society*. New York: Kugler.
Fujimura, J. (1996) *Crafting Science: a Sociohistory of the Quest for the Genetics of Cancer*. Cambridge, MA: Harvard University Press.
Gaudillière, J.-P. (1997) Whose work shall we trust? Genetics, pediatrics and hereditary diseases in postwar France. In Sloan, P.R. (ed.) *Controlling our Destinies. Historical, Philosophical, Ethical, and Theological Perspectives on the Human Genome Project*. Notre-Dame: University of Notre-Dame Press.
Gelehrter, T.D. (1983) The family history and genetic counseling: tools for preventing and managing inherited disorders, *Postgraduate Medicine*, 73, 119–26.
Gelehrter, T.D. and Collins, F.S. (1990) *Principles of Medical Genetics*. Baltimore: Williams and Wilkins.
Gray, J.M. and Conneally, P.M. (1993) Case study of Huntington's Disease. In Frankel, M.S. and Teich, A.H. (eds) *Ethical and Legal Issues in Pedigree Research: Report on a Conference Sponsored by the AAAS Committee on Scientific Freedom and Responsibility and the AAAS-ABA National Conference of Lawyers and Scientists*. Washington: American Association for the Advancement of Science.
Harper, P.S. (1993) *Practical Genetic Counselling*. Oxford: Butterworth-Heinemann.
Howell, J.D. (1995) *Technology in the Hospital. Transforming Patient Care in the Early Twentieth Century*. Baltimore: The Johns Hopkins University Press.
Ishikiri, Y. and Niikawa, N. (1984) Fragile X syndrome, *Knowledge of Medicine (Kusuri no chisiki)*, 34, 7, 46.
Jones, S. (1993) *The Language of Genes: Biology, History and the Evolutionary Future*. London: Harper Collins.
Kanazawa, I. (1994) Diagnosis and counseling for Huntington's Disease. In Fujiki, N. and Macer, D.R.J. (eds) *Intractable Neurological Disorders, Human Genome Research and Society. Proceedings of the Third International Bioethics Seminar in Fukui, Japan, 19–21 November, 1993*. Tsukuba: Eubios Ethics Institute.
Kawashima, H. (1992) Different client's response at genetic clinics in Japan and the USA, and its ethical background. In Fujiki, N. and Macer, D.R.J. (eds) *Proceedings of the Second International Bioethics Seminar in Fukui, 20–21 March, 1992*. Tsukuba: Eubios Ethics Institute.
Keating, P. and Cambrosio, A. (1997) Interlaboratory life: regulating flow cytometry. In: Gaudillière, J.-P., Löwy, I. and Pestre, D. (eds) *The Invisible Industrialist: Manufacturers and the Construction of Scientific Knowledge*. London: Macmillan.
Kessler, S. (1980) The psychological paradigm shift in genetic counseling. *Social Biology*, 27, 167–85.
Latour, B. (1990) Drawing things together. In Lynch, M. and Woolgar, S. (eds) *Representation in Scientific Practice*. Cambridge, MA: MIT Press.
Latour, B. (1994) Une sociologie sans objet? Remarques sur l'interobjectivité, *Sociologie du Travail*, 4, 587–608.

Markel, H. (1992) The stigma of disease: implication of genetic screening, *The American Journal of Medicine*, 93, 209–15.

Marteau, T. and Richards, M. (eds) (1996) *The Troubled Helix. Social and Psychological Implications of the New Human Genetics*. Cambridge: Cambridge University Press.

Mishler, E. (1984) *The Discourse of Medicine: Dialectics of Medical Interviews*. Norwood, NJ: Ablex.

Nakayama, K. (1990) The iconic bottleneck and the tenuous link between early visual processing and perception. In Colin, B. (ed) *Vision: Coding and Efficacy*. Cambridge, UK: Cambridge University Press.

Nelkin, D. (1989) *Dangerous Diagnostics: the Social Power of Biological Information*. New York: Basic Books.

Ohkura, K. (1989) Genetic counselling and ethics, *The Study Meeting of Local Genetic Counselling For Nurses (Kangoshoku no tame no Chiki-idensôdan-kenkyû kai)*, 10, 7–16.

Ohkura, K. and Kimura, R. (1989) Ethics and medical genetics in Japan. In Wert, D.C. and Fletcher, J.C. (eds) *Ethics and Human Genetics: a Cross-Cultural Perspective*. New York: Springer-Verlag.

Olson, D.R. (1996) *The World on Paper. The Conceptual and Cognitive Implications of Writing and Reading*. Cambridge, UK: Cambridge University Press.

Palladino, P. (1997) From family pedigrees to molecular markers: on cancer and heredity at St. Mark's Hospital, 1924–1995. In Gaudillière, J.-P. and Löwy, I. (eds) *Human Pathologies Between Heredity and Infection. Historical Approaches*. London: Harwood.

Poers, M. (1993) Publication-related risks to privacy: ethical implications of pedigree studies, *IRB, A Review of Human Subject Research*, 15, 7–11.

Rabinow, P. (1992) Artificiality and enlightenment: from sociobiology to biosociality. In Crary, J. and Kwinter, S. (eds) *Incorporations*. New York: Urzone Books.

Rabinow, P. (1996) *Making PCR. A Story of Biotechnology*. Chicago: The University of Chicago Press.

Raffel, S. (1979) *Matters of Fact*. London: Routledge and Kegan Paul.

Rapp, R. (1988) Chromosomes and communication: the discourse of genetic counseling, *Medical Anthropology Quarterly*, 2, 143–57.

Rapp, R. (1995) Risky business: genetic counseling in a shifting world. In Schneider, J. and Rapp, R. (ed.) *Articulating Hidden Histories: Exploring the Influence of Eric R. Wolf*. Berkeley: University of California Press.

Resta, R.G. (1993) The crane's foot: the rise of the pedigree in human genetics, *Journal of Genetic Counseling*, 2, 235–60.

Resta, R. and Wcislo, K. (1993) Pedigree software: can it meet counselors' needs? *Perspectives in Genetic Counseling*, 15, 2, 9.

Rheinberger, H.-J. (1995) Beyond nature and culture: a note on medicine in the age of molecular biology, *Science in Context*, 8, 249–63.

Richards, M.P.M. (1993) The new genetics: some issues for social scientists, *Sociology of Health and Illness*, 15, 567–86.

Richards, M.P.M. (1996) Family, kinship and genetics. In Marteau, T. and Richards, M. (ed) *The Troubled Helix. Social and Psychological Implications of the New Human Genetics*. Cambridge: Cambridge University Press.

Rushton, A.R. (1994) *Genetics and Medicine in the United States, 1800 to 1922*. Baltimore: Johns Hopkins University Press.

Saito, Y. (1992) Genetic counselling in Japan (Nihon no Iden Sôdan), *Seirinken Review and Research*, 2, 23–44.

Simpson, S.D. (1993) Bipolar mood disorder. In Frankel, M.S. and Teich, A.H. (eds) *Ethical and Legal Issues in Pedigree Research: Report on a Conference sponsored by the AAAS Committee on Scientific Freedom and Responsibility and the AAAS-ABA National Conference of Lawyers and Scientists*. Washington: American Association for the Advancement of Science.

Star, S.L. (1986) Triangulating clinical and basic research: British localizationists, 1870–1906, *History of Science*, 24, 29–48.

Star, S.L. and Griesemer, J.R. (1989) Institutional ecology, 'translations', and boundary objects: amateurs and professionals in Berkeley's Museum of Vertebrate Zoology, 1907–39, *Social Studies of Science*, 19, 387–420.

Steinhaus, K.A., Bennett, R.L., Resta, R.G., Uhrich, S.B., Doyle, D.L. *et al.* (1995) Inconsistencies in pedigree symbols in human genetics publications: a need for standardization, *American Journal of Medical Genetics*, 56, 291–5.

Strathern, M. (1992) *After Nature: English Kinship in the Late Twentieth Century*. Cambridge: Cambridge University Press.

Suchman, L. (1990) Representing practice in cognitive science. In Lynch, M. and Woolgar, S. (eds) *Representation in Scientific Practice*. Cambridge, MA: MIT Press.

Terrenoire, G. (1986) L'évolution du conseil génétique aux États-Unis de 1940 à 1980: pratique et légitimation, *Sciences Sociales et Santé*, 4, 51–79.

Thévenot, L. (1984) Rules and implements: investments in forms, *Social Science Information*, 23, 1–45.

Thompson, M.W. McInnes, R.R. and Huntington, F.W. (1991) *Thompson & Thompson Genetics in Medicine. Fifth Edition*. Philadelphia: W.B. Saunders Company.

Uchida, H. (1987) How to transcribe a family pedigree: its significance and purpose (Kakeizu no Kisaihôhô – Sono Igi to Mokuteki), *The Study Meeting of Local Genetic Counselling For Nurses (Kangoshoku no tame no Chiki-idensôdan-kenkyû kai)*, 8, 41–56.

Wertz, D.C. and Fletcher, J.C. (1989) *Ethics and Human Genetics: A Cross Cultural Perspective*. New York: Springer-Verlag.

Wood, D. (1992) *The Power of Maps*. New York: Guilford.

Yoxen, E.J. (1982). Constructing genetic diseases. In Wright, P.W.G. and Treacher, A. (eds) (1982) *The Problem of Medical Knowledge. Examining the Social Construction of Medicine*. Edinburgh: Edinburgh University Press.

2. Science versus care: physicians, nurses, and the dilemma of clinical research

Mary-Rose Mueller

Introduction

Nearly forty years ago, Renee Fox (1959/1974) called attention to what was then a nascent domain of modern medical practice, that of clinical research. Drawing on a participant-observation study of a hospital research unit, Ward F-Second, Fox observed that clinical research blurs the boundaries of medical science and medical practice. It requires that physician-investigators experiment with unproven therapies to advance knowledge of human health and infirmity while simultaneously attending to the real life exigencies of patient illness and suffering. Fox documents the difficulties experienced by clinical researchers in 'resolving points of conflict between research and therapy' (Fox 1959/1974: 28) and in ascertaining the limits of their dual roles and responsibilities as investigators and clinicians.

In this chapter, I revisit the enterprise and 'points of conflict between research and therapy' through a case study analysis of AIDS-treatment science in the United States. This analysis, however, differs from Fox's (1959/1974) study in two ways.

First, while Fox focuses on the difficulties of patients and investigators, patients are not directly featured in the present study. Instead, it considers dilemmas that occur between physicians and nurses. Recent increases in public and private support for biomedical science in the U.S. has led to the development of novel organisational arrangements and career paths for professional workers within academic medicine (Hafferty and Light 1995). Previously, clinical research took place on hospital wards, like F-Second, and physicians assumed an active, albeit conflict-laden role, in the actual conduct of experimental procedures. Today, academic medical centres not only have dedicated hospital wards for the conduct of clinical science, but they have also set up specialised outpatient research units. A whole new division of medical labour has emerged to accommodate the expansion of clinical science into these specialised settings. In some instances, physician-investigators attend almost exclusively to the knowledge or 'backstage' work of clinical science: they formulate research problems and protocols, solicit research funding, and evaluate and disseminate research findings (Gray 1975). The actual conduct of clinical trials – executing and coordinating investigatory activities – is frequently discharged by nurses. Indeed, the role of nurses as clinical trial coordinators has been well described in nursing and medical journals (see Eaton and Pratt 1990, Mullin *et al.* 1984). To date, however, the decoupling of physician-investigators from the clinical research enterprise and the movement of nurses into the domain of clinical research has received little notice from medical sociologists. A notable exception is Oakley's (1990) account of the conflicts midwives endure while conducting their own research. Nevertheless, the inattentiveness to nurses in studies of medical research seems somewhat surprising, given that there is some recognition of the importance of 'technicians' in social studies of science (see Mukerji 1989, Shapin 1989).

Second, the analysis presented in this chapter differs from Fox's (1959/1974) study of Ward F-Second in that it places the dilemmas that occur between physicians and nurses within the wider context and organisation of clinical science. To be sure, Fox attends to the larger arena of medical science, linking investigations undertaken on Ward F-Second to broader therapeutic and surgical advancements in the field. Yet Fox locates work-related conflict squarely within the norms and values of non-research-based medical practice and the contextual circumstances of Ward F-Second. In so doing, she eloquently reveals that physician-investigators recognise and are troubled by the clinical limitations of experimental treatments. One way investigators respond to the deficiencies of clinical research is by altering the usual hierarchical and authoritative structure of medical practice: they approach patients as 'pseudo-colleagues' and friends and direct nurses to give patients and families 'red carpet' treatment. The understanding that emerges from Fox's study is that the work of clinical research undermines the normative roles and structure of physician-patient relations and as such leads to the institutionalisation of new norms and values to guide medical care practice.

While there can be little doubt that clinical research challenges the normative order of medical practice and creates dilemmas for physician-investigators, other factors may contribute as well. Here, it is instructive to consider the work of scholars exploring the influence of organisational and contextual forces on the content and practice of medical work and professional-patient relations (Abbott 1988, Anspach 1993, Chambliss 1996, Cicourel 1986, Freidson 1970/1988, Strauss *et al.* 1985, Zussman 1992). Though empirically and theoretically diverse, these scholars seem to agree on at least two issues: first, that the local organisation and wider context of health care shapes and reshapes the structure and relations of work-related undertakings; and second, that efforts to apprehend the link between the organisational context and professional practice must focus on the micro-processes that take place within a given setting, or locale, of work. It must be emphasised, however, that these scholars neither suggest nor assert that the organisational context wholly determines the meanings, practices, and relations of work. Rather, this scholarship seems to suggest that the organisational context of work can be viewed as what Mishler *et al.* (1981/1987:

99) conceive of as a 'context of practice'. Scholarship seems further to suggest that it is within the ecological arrangement of practice that different roles, resources, and obligations are accorded to work participants; it is also where the range of actions by work participants are set forth, employed, and transformed.

I draw on these scholarly insights to examine 'points of conflict' of clinical science, what I term the 'science/care dilemma', and the processes surrounding its depiction and attempted resolution by physicians and nurses within one 'context of practice', that of AIDS-treatment research. Before doing so, I briefly review the research study and setting that forms the basis of this analysis. I then discuss dimensions of the 'science/care dilemma' and show the ways in which this dilemma is affected by the broader context and local organisation of AIDS research.

Study and method

The data were gathered during a field study that took place between December 1990 and September 1991, as part of a larger investigation on the development and expansion of a new jurisdiction of medical practice, that of a multi-site, federally-funded treatment research programme for AIDS (Mueller 1995). The field study took place at one of the federally sponsored sites, or 'centres', for AIDS research (described in detail below). Access to the field site was negotiated through a personal acquaintance who was one of the centre's principal investigators.

The research was initially conceived as an exploratory participant-observation study, the broad objective of which was to witness the naturally occurring processes and practices to be found at the AIDS-treatment research centre. At the beginning of the study, I observed and recorded fieldnotes on a variety of activities, including clinical encounters involving patients and professionals. Shortly after the project was underway, however, I narrowed the focus of inquiry to the interactions and relations of two key professional groups engaged in treatment research work: physician-investigators and trial nurse-coordinators. This decision was driven by both practical and theoretical concerns. On the other hand, I came to understand that clinical research is a highly complex social phenomenon, involving numerous social groups (*e.g.* patients, nurses, physicians) and numerous work practices (*e.g.* recruitment, eliciting informed consent, data management). Temporal considerations required that I limit the groups and activities to be studied in the field. On the other hand, because of the theoretical focus of the larger project, I reasoned that the field experience permitted an insider's view of the interactional and intraprofessional dimensions of the jurisdiction of treatment science in a single practice setting. A narrowed focus on professional interactions and interrelations made it possible to accomplish the field study in a reasonable amount of time and to explore some of the theoretical issues relevant to the larger project.

I renegotiated the objectives and data collection techniques (to include both observations and depth interviews) of the field project with key members of centre staff. I sought out and was granted access to centre events – regularly scheduled staff meetings, AIDS morbidity and mortality reviews, patient conferences, and community advisory board meetings – where physician-investigators and trial nurse-coordinators were likely to interact and talk about the work of clinical research. While attending these meetings, I took hand-written notes of unfolding interactional encounters. Notes were also made of exchanges that occurred between physicians and nurses outside formal meetings. As soon as possible after leaving the setting, I constructed elaborate written accounts from my field notes.

During the course of the field study, there were times when the participants seemed to be keenly aware of my presence. For example, while attending formal meetings between physician-investigators and trial nurse-coordinators, I was occasionally advised that certain information on a particular trial or patient was to be kept confidential. I responded to such warnings with assurances that proprietary information would neither be recorded nor referred to in discussions and reports of the study. About halfway through the field study a trial nurse-coordinator asked if I hadn't 'figured us out yet'. That the physicians and nurses were aware of and expressed interest in my role as field researcher is not entirely surprising. After all, the social complexities of field research for the observer and the observed have been extensively documented (see Emerson *et al.* 1995, Lofland and Lofland 1995). And although it is difficult to know whether my observations disrupted the flow of interactional exchange in any way, I was never asked to leave a meeting nor was I ever precluded from participating in centre events.

Coding-based techniques (see Emerson *et al.* 1995, Lofland and Lofland 1995) were employed to make sense of observations and exchanges recorded in fieldnotes. I then developed an interview guide to explore some of the issues and themes that emerged from observations and analysis of fieldnotes. In particular, I pursued some of the 'points of conflict' over clinical trial work and intraprofessional relations that recurred in discussions between investigators and trial nurse-coordinators. I conducted 19 audiotaped interviews with centre nurses and physicians, transcribing 15 of them verbatim. Occasionally, respondents would reveal sensitive information on their views of, or relations with, other centre staff, and later ask that I keep such revelations 'off record'. All such requests were duly honoured. Interview materials were then coded and analysed in light of coded field notes. Categories and themes developed from field records and interviews were used to structure the analysis presented below.

The organisational context of clinical research for AIDS

The term 'clinical trial work' as it is used here refers to the organised meanings, practices, and relations that constitute the investigatory study of newly developed therapies. A clinical drug trial is a methodologically rigorous approach to human experimentation in which the safety and effectiveness of two or more drugs are assessed and compared, under pre-established and highly specifiable protocol-driven conditions. In the United States, information obtained from clinical trials is used to determine whether or not a drug gets approved and licensed by the Food and Drug Administration; it is also used to determine how a drug will be applied in clinical practice (Lilienfeld 1982).

In 1986, Congress awarded funds to one of the agencies within the National Institutes of Health, the National Institute of Allergy and Infectious Diseases, to institute a multi-site programme of clinical research for AIDS. This programme, the AIDS Clinical Trials Group (ACTG), is charged with conducting controlled clinical trials of promising but unproven agents for AIDS. Since its inception, though, the ACTG has been the object of intense scrutiny by the highly visible and politically astute AIDS activist community as well as the subject of numerous Congressional hearings, a Presidential Commission, and various medical scientific reviews (see Arno and Feiden 1992). Public oversight of the ACTG has taken place, in large measure, because of the high rates of mortality associated with AIDS, the paucity of effective therapies, and the organised efforts of activist groups. Indeed, people with AIDS have aggressively pursued enrollment onto ACTG studies and have often been less than truthful with investigators and nurses about their past medical history (Hendricksen 1988, Weitz 1991). And although the scientific mission, activities, and policies of the ACTG have been established and transformed 'inside of a large and often floodlit arena' (Epstein 1995: 408), the AIDS-affected community has lobbied Congress to expand the number of ACTG studies and to institute practices to increase patient access to trial protocols. Consequently, the ACTG's budget and scope of operations has grown considerably over the past decade (NIAID 1995).

The centre observed here is one of the more than four dozen federally funded sites that make up the ACTG programme. It is also supported with research grants awarded by the state government. The centre's research-related activities are guided by and contingent upon funds received from both the state and federal government. This ACTG centre is located in a county of over two million people. It is affiliated with a local university-based medical school that is acknowledged by the health care community to be a seat of expertise in AIDS-related research, education, and service.

The centre exists as a setting for the conduct of AIDS therapeutic research, and not for the provision of general medical care. In most cases, people with AIDS are referred to the centre by their primary medical doctor for the exclusive purpose of participating in clinical trials. The centre's particular organisational arrangement is congruent with the mission of the ACTG, which is to facilitate clinical research rather than to support the delivery of general health care to persons with AIDS (Institute of Medicine 1991: 26). It is at this centre then that people with AIDS are evaluated for suitability to participate in therapeutic trials and, once enrolled on protocols, receive study-related care. In the early 1990s, when this study took place, the centre was engaged in dozens of clinical trials, involving more than 500 patient-visits each month.

The centre's 'core' staff consist of physician-investigators and trial nurse-coordinators. Just as the centre has been organised for the accomplishment of clinical trial research, the activities of the centre's physicians and nurses have been bureaucratically arranged and hierarchically organised in such a way as to accomplish this practice. At the head of the centre's hierarchy are the physicians. Though all of the investigators are on the faculty at the medical school, the extent of their involvement in affairs outside the centre varies considerably. Three of the centre's six physicians are viewed as national 'experts' in HIV-related clinical, biological, and therapeutic matters. They have played highly visible roles in HIV research at the national level and have been designated principal investigator or co-principal investigators on ACTG and state-sponsored studies. Though they are the most knowledgeable about the biology of AIDS and clinical trial methodology, these physicians have little involvement in the actual operation of centre trials. Senior investigators spend the least amount of time with patient-volunteers and do little to assure the execution of trial protocols. Instead, much of their attention is focused on university-driven research, teaching, and clinical obligations. The senior investigators' engagement with trial-related activities is limited to the 'backstage', or non-clinical, aspects of clinical trial work, like writing grants, designing trial protocols, analysing protocol data, writing papers for publication, and maintaining relations with local physicians. As one investigator put it, the centre's organisation allows 'flexibility for us to pursue other interests, and adds more expertise to the studies'. Only one of the senior researchers has an office at the centre.

There are three other physician-investigators associated with the centre, all of whom have offices on-site. However, these physicians are far less experienced in AIDS research and have far fewer academic responsibilities than their senior colleagues. Nonetheless, these novice investigators have little direct involvement with trial protocols and patient-volunteers. Instead, much of their time is devoted to the backstage work of clinical trials as well as to troubleshooting administrative problems and consulting with nurses on protocol-related concerns.

Nurses constitute the largest portion of the centre's core staff. Seventeen nurses, three of whom are male, are employed in either full- or part-time capacities. The educational experiences of the nurses vary, from licensed vocational training to masters level preparation as nurse-practitioner. Trial nurse-coordinators are responsible for the execution and implementation of the treatment studies: they are actively involved in patient recruitment activities; they oversee the collection of clinical data; and they monitor the progress of patients from the time of enrollment to the termination of the protocol. This means that nurses, rather than the more senior and novice physician-investigators, are directly engaged with trial protocols. It also means that they have the most frequent and sustained contact with patient-volunteers. In addition to protocol activities, trial nurse-coordinators attend to various administrative and educational duties.

In sum, AIDS-treatment research takes place within a politically charged environment. At the local level, it is organised to accomplish key features of the work, including the receipt of grants, the analysis and dissemination of research findings, and the recruitment to and maintenance of patients on study protocols. As we shall see, the organisational context of clinical research influences and shapes the dilemmas involved in this line of work. Before discussing the relationship between the context of practice, and work-related conflict, I review dimensions of the science/care dilemma, as it was expressed by the centre's physician-investigators and trial nurse-coordinators.

The science/care dilemma and its expression

Conflicting views of patient-volunteers

One expression of the science/care dilemma relates to the status of patient-volunteers. Patients are crucial to the successful accomplishment of clinical trials. Before participating in a clinical trial, however, patients have to be able to meet the inclusion criteria specified in the protocol. For example, they must have laboratory or radiographic evidence of the disease under consideration, fall within a certain age range, or be free of confounding medical conditions. Patient-volunteers must also have the ability to comply with the terms and conditions set forth in the protocol. For example, they must be able to take study medications as prescribed, to keep appointments with trial nurse-coordinators, and to obtain diagnostic evaluations as required.

But perhaps more importantly, the participation of patient-volunteers on clinical trials is essential to the careers of physician-investigators. As previously noted, academic medical careers in clinical research are supported to a large extent with grants from governmental agencies. Grants are legal contracts that outline the undertakings to be performed and supported within a given period of time; as well as being time-limited they are also competitively awarded. As such, physician-investigators must arrange the local practice setting in such a way as to facilitate the successful discharge of their contractual obligations, as well as to facilitate the advancement of their careers within the academic environment.

The centre observed here is primarily supported with a grant from the National Institute of Allergy and Infectious Diseases under the auspices of its AIDS Clinical Trials Group, or ACTG, programme. Consequently, the centre is obliged to enroll large numbers of patients onto ACTG-sponsored clinical trials. One of the challenges confronting centre investigators is the identification of potential patient-participants from the finite number of AIDS-affected persons in the region and, once identified, of enrolling them onto clinical trials. To ensure ready access to patients for their studies, physician-investigators have established referral networks with medical providers caring for AIDS-affected persons. Another challenge faced by centre investigators is conforming to the explicitly-stated purpose of the grant, which is the implementation of clinical trials. To meet this challenge, physician-investigators have oriented centre activities toward clinical trial care, rather than the provision of general medical services. Investigators have also arranged the centre to meet the demands of their academic responsibilities and backstage trial obligations. As such, nurses have been hired and put in charge of the day-to-day operations of trial protocols.

In the course of conducting this field study, it became apparent that physician-investigators and trial nurse-coordinators hold very different views of the patients who volunteered to participate in the centre's studies. On the one hand, physicians have a somewhat narrowly drawn understanding of individual trial participants. For them, patients are seen as subjects of medical experimentation, and are valued for their contribution to broadly stated goals of clinical science and the grant-defined mission of the centre rather than for their peculiar personal qualities or medical infirmity. That centre investigators view patients in this way is not too surprising. Indeed, the objectification of human subjects is the focus of much of the discourse on the ethics of clinical trials. As one one physician-ethicist put it, clinical research

> transforms the doctor-patient relationship into something other than the fiduciary relationship in which the physician's sole concern is the well-being of the individual patient. Amelioration of that conflict of interest is the real goal of ethical guidelines [for research] (Lantos 1993: 2814).

On the other hand, centre nurses have a very different view of patients than the investigators. They conceive of patient-volunteers as normal individuals, albeit afflicted and infirm, and value them in part because of their contribution to medical science and in part because of their dependence on centre staff to look out for their medical care interests.

This distinction is illustrated in a discussion that ensued during one of the centre's regularly scheduled staff meetings, a forum wherein problematical issues – new and on-going – are presented and debated. At issue were the medical records of patient-volunteers. The chief nurse-administrator expressed concern that a medical researcher from outside the centre was photocopying information from patient charts. She asserted that this use of the centre's documents represented a breach of patient confidentiality and a misuse of patient trust. The physicians, however, had a very different response to the situation. One investigator asked if there was any benefit in the arrangement, if the centre's staff would be invited to participate on research grants generated from the data, in exchange for the release of patient information. A senior physician-investigator informed the group that the sharing of patient data was 'an outgrowth of collaboration [with other medical school researchers] from a very long time ago', but that a proviso had been instituted to guarantee that the centre would be involved in and credited with studies that came out of the arrangement.

Another point of disagreement over the rights of patients emerged around the issue of patient withdrawal from on-going clinical trials. This issue was taken up at a staff meeting, in which one of the nurses informed the investigators that some patients wanted to drop out of an established study and enroll onto a recently instituted protocol designed to evaluate the effectiveness of a new anti-AIDS agent. She asserted that patients believed in the superiority of the new drug and protocol. The nurse then asked investigators if there was any scientific justification for the existing study, and if there was not, she wanted to know 'why should we [nurses] encourage them to stay on'? One investigator responded that there were two factors to consider: one was 'what the patients want', and the other was 'what we [the researchers] don't want', which was for patients to drop out of the study prematurely, before all the data could be analysed. Another investigator added that 'we don't know what's working [therapeutically for AIDS] and what isn't. If we did, we'd put [patients] on it'. Investigators then advised the nurses to encourage patients not to drop out of the older study, and to follow the protocol to its natural completion.

Physician-investigators and trial nurse-coordinators either implicitly or explicitly disagree over expectations for patient involvement on clinical trials as well. As previously stated, physicians develop clinical protocols to answer questions about new or previously unproven therapies. But not all protocols are structured in the same way. In fact, there is tremendous variability in protocol design, especially as it relates to such issues as the duration, intensity, and utilisation of human and economic resources required to answer clinically relevant questions. Not unexpectedly, some protocols are more labour intensive than others, and some protocols are more demanding of patient-volunteers. Yet as has been observed by AIDS activists and advocates (see Epstein 1996), the effects of protocol requirements on the lives of patients is rarely taken into account in the development and design of clinical trials. This observation has also been made by trial nurse-coordinators.

During an interview in which various aspects of trial clinical work were explored, a nurse described problems recruiting patients onto an intensive, two-week clinical trial. Not only did she have a hard time identifying patients with the stamina to endure the terms of the protocol, but once she did, it was difficult to convince them that their time was worthy of the research effort:

> I try to encourage people . . . Some people say I only have a two week vacation and I really think I ought to go see my mother. And I tell them they have to weigh the cost and benefit of this. And I really feel like, if it were me and I had to choose two weeks doing a study . . . or doing something that was really important to my life, then I would do that.

Investigators and nurses also express disagreement on the limits of patient participation after studies had been terminated. During a staff meeting, a discussion focused on whether data should continue to be collected on patients for the centre's observational data base after they had completed a clinical trial. The nurses expressed concern over the utility of maintaining patient involvement in non-protocol-related centre activities. For investigators, however, this practice was a means for acquiring additional information on disease progression, data that one investigator said were essential 'for the long-term follow up' of disease management.

The conflicts presented above – sharing protocol data and maintaining patient adherence to on-going studies – reflect how the nurses and investigators view the moral status of patient-volunteers. In the dilemma over medical records, nurses held that patients are imbued with citizenship rights to privacy and, at the very least, the exercise of this right means that they should be consulted before information from their medical records is made available to non-centre researchers. However, physicians held that medical records are not to be confused with individual patients. For them, records are objectifications or representations of physiological processes, and not actual patients. Investigators further held that because records are assembled in the course of conducting research, data belonged to the researchers and could therefore be used to advance the goals of clinical science without recourse to patient-volunteers. In disagreements over patient participation, nurses championed the right of patients to decide when, and if, to withdraw from protocols. Investigators emphasised that while there may be little scientific benefit to remaining on studies for individual patients, there may be benefits accrued to many patients in the future. Thus, facets of the science/care dilemma are tightly linked to the status attributed to patient-volunteers: physicians consider patients as extensions of the experimental project whose lives may be reshaped to fit the goals, contingencies, and demands of clinical science; nurses regard patients from the vantage point of the context and complexities of their lives as infirm persons with rights to control their medical care activities. As will be discussed later, these divergent views of patient-volunteers are connected to the socially organised roles of physician-investigators and trial nurse-coordinators.

Conflicting boundaries of clinical trial practices
Just as there are differences in the way investigators and nurses view patient-volunteers, there are also differences in the way they perceive the practice boundaries of clinical trial work. Clinical trial protocols are, at least in the ideal, templates for action: they define the terms and conditions of the investigation and, by extension, they guide the activities and practices of clinical trial workers. These activities are at least in part grant-supported. Indeed, grants cover the direct costs incurred in the conduct of clinical trials, such as laboratory and radiographic tests, pharmaceutical materials, and salaries of trial personnel. Grants also fund indirect costs, like organisational overhead and unanticipated patient care expenses not directly specified in clinical trial protocols.

Grant-funded clinical trial work is a form of medical practice that is geared toward the production of knowledge on the effects that specific therapeutic interventions have on specific disease conditions. Patient-volunteers are recruited onto clinical trials precisely because they are medically infirm. In the case observed here, patient-volunteers are infected with HIV and, in some instances, suffer from some of the devastating and oftentimes fatal conditions associated with AIDS. During an interview to explore the clinical dimensions of AIDS, one nurse commented that people with AIDS are 'acutely ill with a life-threatening infection, and they're very ill, and they require a lot of effort and they require a lot of monitoring . . .'

As this quote indicates, patient-volunteers require continuous medical attention. This requirement, however, often extends into the realm of general medical care, beyond that which is specified in and funded by the centre's clinical trial grants. But many of the centre's patient-volunteers are uninsured, or underinsured, and therefore are precluded from accessing medical care providers and general health care services. Although physician-investigators and trial nurse coordinators are aware of this problem, they disagree on the extent to which clinical trials should be used to accommodate the unmet health needs of patients. On the one hand, physician-investigators view the work of clinical trials as a scientific enterprise and as such, seek to draw the boundaries around clinical trial practice narrowly, in effect, to limit the delivery of non-study care delivered at the centre. On the other hand, nurses view clinical trials as an extension of medical care, complimenting the treatment needs of patient-volunteers. Not infrequently, nurses seek to expand the practices of clinical trial work into the extra-study domain of general health care.

The expression of this dilemma, and the view of physician-investigators, was brought into clear relief during staff meetings. For example, at one meeting in which protocols for opportunistic infections were being reviewed, an investigator enumerated the diagnostic and therapeutic care to be rendered under the terms laid out in the study's grant. He remarked that extra-study care was to be coordinated with the patient's primary care physician and ultimately billed to the patient's insurance company. A trial nurse informed the physician that some of the patients willing to participate on these studies were uninsured, and that this presented the problem of finding medical doctors to oversee their general health care and the problem of financing their extra-study care. One investigator responded that 'subjects without insurance may have to be restricted from [these] studies'. This view was further underscored at a staff meeting the following week, in which an investigator declared 'we're here to provide protocol care. The amount of clinical care is supposed to be minimal'. On the occasion of another staff meeting, a trial nurse suggested that because of the press to increase the number of patients enrolled onto studies, the centre should offer preventive health care to men and women, as a value-added benefit of their participation on clinical trials. One of the chief investigators rejected this suggestion, maintaining that health promotion services should be administered in primary health-care settings and not the clinical trial centre. This theme was further elaborated in an interview with an investigator in which the medical care needs of study participants was probed:

> It is very clear, both from the stand point of our sponsoring institutions and our own goals, that we cannot use the scarce resources [of research for primary care]. That would just suck up these resources. There's a lot of clarity in our minds [about this].

Unlike the investigators, though, nurses hold a very different, less narrowly circumscribed, understanding of clinical trial work. Protocols are seen not only as part of the scientific enterprise, but also as an integral component of the patient's medical care. Nurses are charged with the operations and management of all phases of protocol implementation, from recruitment to completion. Consequently, nurses oversee the 'career' of patient-volunteers as they progress on the protocol. In some cases, nurses follow patients over a period of months and years, during which time patients are likely to develop medical complications of HIV infection. This long-term involvement in protocol care, coupled with the changing medical status of patient-volunteers, fosters the development of close relations between nurses and patients. Indeed, nurses often describe protocol participants as their 'patients'. One trial nurse-coordinator put it this way:

> Here it's week after week, month after month, year after year . . . You watch them suffer. You break bad news to them, tell them 'you have your third AIDS infection.' You watch their lovers die of AIDS in their arms. You watch their friends reject them. You watch them in pain. You watch them [become] physically disfigured. These are huge, huge issues as far as how can you just do your job.

Nurses not only 'watch' patient-volunteers over the course of their tenure on clinical trials, as this nurse put it, they also become actively involved in their patients' extra-study affairs and medical problems. Not infrequently, patients are acutely ill at the time of protocol visits. And although attending to acute illnesses is not a protocol-defined activity, nurses feel compelled to intervene. In some instances, trial nurse-coordinators are able to contact the patient's health care provider and arrange for immediate attention. In other cases, patients receive general health care at the medical centre, from investigators acting in the discharge of their academic duties as 'clinicians' or 'attendings'. When general health care problems arise for these patients, trial nurses simply arrange for a consultation with the investigator. In still other cases, trial nurses consult with centre investigators, eliciting advice on the management of clinical care problems. During an interview in which dimensions of this dilemma were explored, one nurse recounted a situation in which she was able to discuss the medical disposition of a very ill patient with one of the centre's investigators:

> I was able to bring in the doctor . . . to help me decide is this something we can start a work-up with, or something that should go back [to the general health care physician], or is it, or is it not, drug-related toxicity.

There are situations, however, when nurses take a more active role in the clinical care of study patients. This usually occurs when patients are in need of extra-study medical attention but, because they lack health insurance, do not have access to general medical care. One nurse underscored the difficulty of dealing with this clinical management issue. 'I have some patients that have no primary care and I am providing primary care for them as a nurse with very little physician support. And that is a problem'. The inclination of nurses to breach the boundaries of clinical trial protocols is not entirely unexpected. Oakley (1990) has reported, for example, that midwives tinker with the randomisation process to insure that patients receive interventions best suited to their clinical needs.

The tendency of nurses to want to intervene in the non-study-related medical care problems of patient-volunteers has not escaped the attention of the centre's physicians. One investigator claimed that the impulse displayed by nurses 'is often to get more involved . . . they constantly fret about whether they're appropriately addressing the patients' needs versus exploiting them for research'. As will be seen in the next section, this has been an on-going problem for centre participants and, as a result, physician-investigators and trial nurse-coordinators have sought to come to terms with the science/care dilemma.

The science/care dilemma and its management

Reorganising roles and practices

As previously mentioned, centre participants are aware that they disagree on matters of patient-volunteers and the clinical trial enterprise. Nurses recognise that it is more difficult for them than for investigators to keep their roles as researchers and caregivers distinct from one another. Investigators recognise that, unlike the nurses, they are able to segment their roles and give priority to research over patient care activities. One investigator illustrated this role segmentation during an interview in which he discussed the case of a recently hospitalised, non-study patient under his care.

> Right now I could very easily be visiting a patient [in the hospital] that has an acute problem that may or may not need surgery. But I'll see him later because I think the thought processes that are going on here [at the centre] are important to my [commitment] to research.

One of the strategies that had been institutionalised to address these differences, as well as other pressing issues and concerns to the investigators and nurses, was the weekly staff meeting. Over time, however, the weekly staff meeting was perceived by centre participants as an inadequate forum for handling such concerns. As such, investigators hired a team of consultants to assess the issues contributing to intraprofessional stress and conflict and, on the basis of that assessment, to develop a plan for reducing or obliterating it. After interviewing centre staff, the consultants determined that there was widespread disagreement over both the proper execution of clinical trial protocols and the proper execution of professional roles and responsibilities. Consultants recommended, among other things, that the roles of physicians and nurses be clarified and more formally explicated.[1] In the months following this recommendation, investigators devised and instituted policies to manage the relations and practices of centre participants better. Although informal policies had previously been in existence, the implementation of new policies and procedures represented an attempt to reduce the conflicts experienced by centre participants.

One set of policies reordered the centre's organisational structure. Two new positions – Clinical Director and Administrative Director – were created and accorded to two of the less senior investigators. In addition, the role obligations of the Nurse Administrator were restated and fortified. Individuals in these positions have been given formal authority over various administrative duties and tasks within the centre. For example, the Clinical Director is to oversee the implementation of standard operating procedures, infection control policies, and issues of patient confidentiality. The Administrative Director is to coordinate centre and non-centre activities of physician-investigators, to administer the utilisation of human and fiscal resources, and to act as the centre's liaison with the university and other community organisations. The Nursing Director is to supervise all nursing activities, to liaise between the centre's physicians and nurses, and to assess and facilitate the educational needs of the nursing staff.[2]

The second set of policies, the Clinical Practice Guidelines, are oriented toward the centre's clinical trial activities. These Guidelines acknowledge that non-study medical problems occur in the course of conducting clinical trials. As stated in the preface to the policy: 'The primary role of the [Center] is to conduct clinical research trials. A degree of clinical care may be needed in the course of such trials'.[3] In light of this recognition, the Guidelines set out a plan of action. All protocols are to be assigned a 'Protocol Physician', usually is the 'primary' investigator of record for the study. Among other duties, protocol physicians are to supervise the implementation of the trials under their responsibility, collaborate with trial nurses on study-related care and, if necessary, coordinate all extra-study medical care with the primary physicians of patient-volunteers. Protocol physicians are to 'review primary care coverage with research nurse on regularly scheduled basis . . . and arrange communication with primary care providers if needed'.[4] In the event that the Protocol Physician is unavailable, however, the Guidelines detail the procedures to be followed. For example, at least one investigator is to be present at the centre at all times. This investigator, or 'Doctor of the Day', is to take over the responsibilities of the Protocol Physician, and as such evaluate urgent clinical care problems and formulate the 'appropriate disposition [of the patient], including transport and follow-up with the primary physician'.[5]

Thus, the intent of these policies is to manage the difficulties that arise in the course of conducting clinical science. On the one hand, they reorder the relations between the investigators and nurses, giving formal authority for the administration of clinical trial work to the physicians. On the other hand, these policies underscore the work of clinical research as a scientific, rather than as a general medical care-giving, enterprise. Consequently, these policies serve to legitimate the views of physician-investigators regarding clinical trial work and patient participation. As we shall see, however, it is possible that, even with the implementation of formal policies and procedures, centre participants may not be unable to escape entirely from the science/care dilemma.

Unresolved dilemmas

Although I left the field site shortly after these two sets of policies had been fully instituted, an event that occurred before my leave-taking suggests the difficulties of managing the science/care dilemma in clinical trial work. During a weekly staff meeting, centre participants discussed the disposition of 'disability forms'. Nurses informed investigators that it was not uncommon for patients to request that centre personnel fill out the forms, documenting their disabilities and their qualifications for general welfare assistance. At issue was whether or not the completion of disability forms represented a protocol or non-protocol related activity. Nurses expressed their willingness to complete the forms, maintaining that doing so was a way of compensating patient-volunteers for participating on trial protocols. They further maintained that because of their long-term involvement with patients on clinical trials, they were well positioned to assess and detail the debilitating and disabling condition of study volunteers. Physician-investigators, however, viewed the completion of disability forms as unrelated to the conduct of clinical trials. One researcher asserted that it was not only time consuming and cumbersome to do, but that the 'danger of doing so creates the impression of [our] being the primary care doctor'. Another investigator agreed with his colleague, and added that 'it'll start us on a slippery slope [. . . on which] we'll owe them hospitalisation [benefits as well]'.

By the end of the meeting, one of the investigators forged a compromise that was agreeable to both groups. If the patient's 'participation in the study depends on it [. . . or] if filling out the form is easy to do with the patient', then it was acceptable for the nurse to do so. However, if the task proved too time consuming and/or required more medical information than was already available at the centre, then the responsibility was to be given over to the patient's general care physician.

Although it is impossible to make definitive predictions for the future, it appears that it may be difficult, even in the wake of reorganisation, fully to resolve the science/care dilemma of clinical research. Consequently, it seems likely that dilemmas of this sort will continue to arise in this setting.

On social organisation and (medical) work-related conflict

The preceding analysis has shown that the major participants in clinical research for AIDS – physician-investigators and trial nurse-coordinators – hold divergent views of the patients and practices involved in this line of work. It has also shown that while participants strive to overcome the dilemmas that surface in the course of conducting their work, it may not be possible entirely to manage or resolve 'points of conflict' in clinical research. What accounts for the science/care dilemma and its probable resistance to alteration?

One possible explanation is that this dilemma is linked to differences in the career preparation and sex role socialisation of physicians and nurses. Undoubtedly, medicine remains a male dominated profession and modern medical education and training stresses the 'scientific' basis of clinical decision-making and practice (Starr 1982). And there is little dispute that nursing is a profession of women and that nursing education continues to focus on the 'caring' elements of health and healing (Reverby 1987).

While opposing ideologies and gendered professional role expectations may account for some of the expression of differences about patients and clinical trial work, it cannot, I believe, account for all of it. Nor can it account for the fact that physician-investigators, and not trial nurse-coordinators, are able to institute formal policies and procedures in an effort to manage work-related conflict. And as we have seen, policies instituted at the centre in effect affirm the views embraced by physician-investigators. That attempts to manage dilemmas were devised in a manner consistent with the views of centre physicians is less an indicator of differences in career trajectories and gender norms than it is a reflection of institutionalised patterns of power and inequality. Fully to apprehend the dilemmas experienced by physicians and nurses, I believe it is necessary to consider the ways in which the broader context and local organisation shapes and influences the roles and relations of clinical research.

The practice of clinical research presented above represents a distinctively modern form of medical work. Indeed, it appears that the organisational arrangement of the ACTG is not unusual. Much of the biomedical research that takes place in the United States is funded by the federal government and, to a lesser extent, by state and local governments. The institutional structure of the ACTG – multi-site, academic medical centre based, research focused – is similar to other federally-funded centres and programmes of research like, for example, the National Cancer Institute's Cancer Treatment and Control Program (Kaluzny *et al.* 1996). Moreover, the hierarchical organisation and separation of research from general clinical care evident in this ACTG has been institutionalised in other clinical trial settings as well.

The separation of clinical research from clinical care as exemplified in the ACTG centre differs significantly from the integrated model seen in Fox's study (1959/1970) and other social science studies of clinical research (Barber 1980, Crane 1977). Indeed, like other facets of health care (Freidson 1994), the work of clinical research has become more highly differentiated. As a result, the practice of clinical research has undergone, to paraphrase Abbott (1988: 151) a 'split of workplace jurisdiction' along professional and technical lines. This split is similar to that found in other facets of the division of health care labour, including medical schools (Cicourel 1986), hospitals (Strauss *et al.* 1985), and mental institutions (Goffman 1959, Scull 1994).

Whereas the physicians of Ward F-Second assumed oversight for and provided some of the study and non-study medical care of patient-volunteers (Fox 1959/1970), the physician-investigators of the AIDS-treatment research centre are, for the most part, removed from the day-to-day management of protocol and clinical care needs of patient-participants. Instead, most of their attention is centred on the 'scientific' aspects of clinical trial work. The technical or 'caring' dimensions of clinical research activities have been delegated to nurses.

The local organisational setting and the broader environment of clinical research for AIDS then forms, to paraphrase Anspach (1987), a 'context of practice', in which occupational groups enact work-related roles and utilise work-related resources. Physician-administrators are hierarchically removed from day-to-day interaction with patient-volunteers. The structured distance of physician-investigators from patients and the actual work of clinical research shapes their views of the moral status of patients and the boundaries of clinical trial practice. Moreover, by virtue of their power and status within the medical centre, as well as their authority as grant recipients, physician-investigators have a broader vantage point from which to distinguish the clinical research enterprise from the provision of general medical care. And it is the structure and organisational arrangement of modern medical research that places physician-investigators in the position of having to conform to the trial-centred terms and conditions outlined in research grants.

Indeed, the boundary between general health and grant-funded research care has long been discussed in the medical literature. As one National Cancer Institute researcher put it: 'The actual clinical-care costs [of non-study care] obviously cannot be borne by the government [in the form of clinical research grants]' (Kaufman 1993: 2804). In the case of federally-sponsored research, physician-investigators are accountable to medical administrators for their research and budgetary activities (Mueller 1995). Consequently, investigators have a significant stake in making sure that the resources and requirements of the centre's grants are utilised for research, and not for general health care services.

The organisational arrangement and larger contextual structure of clinical research shapes the perspectives of nurses as well. As discussed above, nurses view study participants as their 'patients' rather than research subjects. As the occupational group in closest proximity to patients, trial nurses are aware of the structural factors that limit access to general medical and research-related care. Indeed, as Weitz (1991) points out, people with AIDS are frequently uninsured or underinsured and therefore participation on clinical research studies often serves as *entrée* to the medical care system. Nurses are confronted with the day-to-day problems that arise in the course of executing trial protocols; they are also confronted with the long-term consequences of a debilitating disease like AIDS for patient-volunteers. Unlike physician-investigators, trial nurse-coordinators have a more narrowly circumscribed, patient-centred perspective on clinical research than physician-investigators. Given their long-term relations with patient-volunteers, nurses have a significant interest in trying to push the limits and boundaries of clinical research to accommodate the study and non-study medical care needs of patients. Yet because of well entrenched structures of power and authority, there are limits to the extent to which nurses can intervene in the non-study care of patients. As we have seen, those limits have been set by research grants, physician-investigators, and formal policies and procedures.

The dilemmas reported above – differing perspectives on the moral status of patients and the boundaries of medical work – appear to resonate with recent scholarship on the effects of organisational context on medical practice. For example, Anspach (1987, 1993) examines decisions that are made to sustain or terminate the medical care of high-risk newborn infants in neonatal intensive care units. She found that physicians, nurses, and (to a lesser extent) parents, use different types of information and different modes of understanding to calculate the prognostic fortune of neonates. Anspach argues that the 'life-and-death' decisions made by these groups are embedded within the organisation of medical practice and the historical context of bioethics. She shows that physicians have a detached relationship with hospitalised infants and rely on technological data to formulate diagnostic and treatment plans. Nurses employ interactional cues, culled from direct caregiving activities, to assess infant survivability. And despite policies to increase patient self-determination in matters of health care, parents have little involvement in medical decisions because their actions are 'managed' by professionals. Zussman (1992) places intraprofessional conflicts that arise in the care of the seriously ill within the context of biomedical ethics and the organisation of hospital intensive care practice. He shows that physicians circumvent national and organisational level policies designed to increase patient and family involvement in health care decisions by redefining 'ethical' problems, like terminating life prolonging medical treatment, into 'technical' or clinical management, problems. As such, physicians are able to maintain decision-making authority over patient care. He further shows that physicians and nurses differ on the meaning of 'intensive care' practice: whereas nurses often see it as 'comfort and care', physicians view it as 'aggressive treatment and cure' (Zussman 1992: 70).

Unlike the work of Anspach (1987, 1993) and Zussman (1992), however, this study was not undertaken to observe the actual practice of clinical research. As mentioned above, its scope was more modest: to explore the interactional and intraprofessional relations of clinical research for AIDS. It may be that prolonged observations in the field would reveal that, despite the broader context of grant-funded AIDS research and the institutionalisation of local policies and procedures, there is a local 'negotiated order' (Strauss *et al.* 1981) of clinical trial work, similar to that observed by Anspach (1987, 1993) and Zussman (1992). Indeed, it would be interesting to observe how the practices of clinical research are actualised *in situ*, as well as how patient-volunteers, trial nurse-coordinators, and physician-investigators define and redefine the boundaries of science and care. Further, it would be interesting to detail how (and if) patient-volunteers are able to influence the decisions of physician-investigators with regard to managing non-protocol, general health matters. Observations of this sort would be especially important to record, given the influence of AIDS activists over federal sponsorship for research (Mueller 1995) and over the design and interpretation of clinical trial research (Epstein 1996). Such a study would make a nice comparison with Anspach (1993) and Zussman (1992) who show that while broader contextual forces have had some influence over the actions of medical doctors, they have not seriously limited their discretion to make patient-care decisions. Indeed what these studies strongly suggest is that even though physicians are under broader contextual constraints, the organisational structure of health care preserves, to a large extent, medical discretion and autonomy over intraprofessional relations and medical practice.

Acknowledgements

I am grateful to the physician and nurse participants in this study. I should like to thank Judith C. Barker, Aaron V. Cicourel, J. Allen McCutchan, Hugh Mehan, Linda S. Mitteness, Diana Torres, Fredric D. Wolinsky, and the editor and referees for their comments on previous drafts of this chapter. The preparation of this chapter was partially supported by National Institute on Aging Training Grant T32AG00045, Linda S. Mitteness, Director.

Notes

1 Untitled centre internal document: undated.
2 Untitled centre internal document: 1991.
3 Clinical Practice Guidelines, centre internal document: 1991.
4 Clinical Practice Guidelines, centre internal document: 1991.
5 Clinical Practice Guidelines, centre internal document: 1991.

References

Abbott, A. (1988) *The System of Professions*. Chicago: University of Chicago Press.

Anspach, R.R. (1987) Prognostic conflict in life-and-death decisions: the organization as an ecology of knowledge, *Journal of Health and Social Behavior*, 28, 215–31.

Anspach, R.R. (1993) *Deciding Who Lives*. Berkeley: University of California Press.

Arno, P.S. and Feiden, K.L. (1992) *Against the Odds*. New York: Harper Collins Publishers.

Barber, B. (1980) *Informed Consent in Medical Therapy and Research*. New Brunswick, NJ: Rutgers University Press.

Chambliss, D.F. (1996) *Beyond Caring: Hospitals, Nurses and the Social Organization of Ethics*. Chicago: University of Chicago Press.

Cicourel, A.V. (1986) The reproduction of objective knowledge: common sense reasoning in medical decision making. In Bohme, G. and Stehr, N. (eds) *The Knowledge Society*. Dordrecht: D. Reidel.

Crane, D. (1977) *The Sanctity of Social Life: Physicians' Treatment of Critically Ill Patients*. New Brunswick, NJ: Transaction Books.

Eaton, R. and Pratt, C.M. (1990) A clinic's perspective on screening, recruitment and data collection, *Statistics in Medicine*, 9, 137–44.

Emerson, R.M., Fretz, R.I. and Shaw, L.L. (1995) *Writing Ethnographic Fieldnotes*. Chicago: University of Chicago Press.

Epstein, S. (1995) The construction of lay expertise: AIDS activism and the forging of credibility in the reform of clinical trials, *Science, Technology and Human Values*, 20, 408–37.

Epstein, S. (1996) *Impure Science: AIDS, Activism, and the Politics of Knowledge*. Berkeley: University of California Press.

Fox, R.C. (1959/1974) *Experiment Perilous*. Philadelphia: University of Pennsylvania Press.

Freidson, E. (1970/1988) *Profession of Medicine*. Chicago: University of Chicago Press.

Freidson, E. (1994) *Professionalism Reborn*. Chicago: University of Chicago Press.

Goffman, E. (1959) *Asylums*. New York: Anchor Books.

Gray, B.H. (1975) *Human Subjects in Medical Experimentation*. New York: John Wiley and Sons.

Hafferty, F.W. and Light, D.E. (1995) Professional dynamics and the changing nature of medical work, *Journal of Health and Social Behavior* (Extra Issue), 132–53.

Hendricksen, C. (1988) The AIDS clinical trials unit experience: clinical research and antiviral treatment, *Nursing Clinics of North America*, 23, 697–706.

Institute of Medicine (1991) *The AIDS Research Program of the National Institutes of Health*. Washington, DC: National Academy Press.

Kaluzny, A.D., Warnecke, R.B. and Associates (1996) *Managing a Health Care Alliance: Improving Community Cancer Care*. San Francisco: Jossey-Bass Publishers.

Kaufman, D. (1993) Cancer therapy and the randomized clinical trial, *Cancer Supplement*, 72, 1801–4.

Lantos, J. (1993) Informed consent: the whole truth for patients? *Cancer Supplement*, 72, 2811–15.

Lilienfeld, Abraham M. (1982) Ceteris paribus: the evolution of the clinical trial, *Bulletin of the History of Medicine*, 56, 1–18.

Lofland, J. and Lofland L.H. (1995) *Analyzing Social Settings: a Guide to Qualitative Research* (3rd Edition). Belmont: Wadsworth Publishing Company.

Mishler, E., AmaraSingham, L.R., Hauwer, S.T., Liem, R., Osherson, S.D. and Waxler, N.E. (1981/1987) *Social Contexts of Health, Illness, and Patient Care*. Cambridge: Cambridge University Press.

Mueller, M.R. (1995) Science in the Community: the Redistribution of Medical Authority in Federally Funded Treatment Research for AIDS. Unpublished Doctoral Dissertation, University of California, San Diego.

Mukerji, C. (1989) *A Fragile Power: Scientists and the State*. Princeton, NJ: Princeton University Press.

Mullin, S.M., Warwick, S., Akers, M., Beecher, P., Helminger, K., Moses, B., Rigby, P.A., Taplin, N.E., Werner, W., Wettach, R. and the MILIS Group (1984) An acute intervention trial: the research nurse coordinator's role, *Controlled Clinical Trials*, 5, 141–56.
National Institute of Allergy and Infections Diseases (1995) NIAID Funds Adult AIDS Clinical Trials Group, *AIDS Agenda*, 1, 14–15.
Oakley, A. (1990) Who's afraid of the randomized controlled trial? Some dilemmas of the scientific method and 'good' research practice, *Women and Health*, 15, 25–59.
Reverby, S.M. (1987) *Ordered to Care*. Cambridge: Cambridge University Press.
Scull, A. (1994) *The Most Solitary of Afflictions*. Princeton, NJ: Princeton University Press.
Shapin, S. (1989) The invisible technician, *American Scientist*, 77, 554–63.
Starr, P. (1982), *The Social Transformation of American Medicine*. New York: Basic Books.
Strauss, A., Schatzman, L., Bucher, R., Ehrlich, D. and Sabshin, M. (1981) *Psychiatric Ideologies and Institutions*. New Brunswick, NJ: Transaction Books.
Strauss, A., Fagerhaugh, S., Suczek, B. and Wiener, C. (1985) *Social Organization of Medical Work*. Chicago: University of Chicago Press.
Weitz, R. (1991) *Life with AIDS*. New Brunswick: Rutgers University Press.
Zussman, R. (1992) *Intensive Care*. Chicago: University of Chicago Press.

3. Bodies of knowledge: lay and biomedical understandings of musculoskeletal disorders

Helen Busby, Gareth Williams and Anne Rogers

Introduction

In this chapter we explore understanding of musculoskeletal disorders among lay people affected by a range of symptoms in muscles and joints, and examine their relationship to the territory dominated by medical science. In a field where the acceptance of a particular stance clearly has implications for policies in health care and beyond, we find a reminder that although the symbolic power of biomedical knowledge is substantial, its clinical effectiveness is less so. Against this background the lived experience of illness can create doubt about received expertise, encouraging the development of a more situated or relational knowledge. (Comaroff 1985)

Musculoskeletal disorders provide an interesting case-study through which to examine these issues. Medical knowledge of these disorders has been characterised by progressively more recondite models of sub-cellular pathogenesis, driven by the concerns of scientists working in genetics, immunology, and biochemistry. Within rheumatological practice, however, treatment even for those inflammatory disorders that have been highly researched, such as rheumatoid arthritis, remains of limited therapeutic efficacy (Williams *et al.* 1996). Moreover, the bulk of symptoms affecting the musculoskeletal system continue to be self-treated or managed in general practice. In this gap between biomedical knowledge and clinical practice, lay knowledge becomes a way of making sense of symptoms in the context of everyday experience (Comaroff and Maguire 1981, Williams 1984).

We discuss the parallels between this process and those involved in the 'public understanding of science', and then go on to describe the main differences between lay and professional understandings of musculoskeletal disorders. We continue by discussing our own study of musculoskeletal symptoms, exploring the relationship between lay and specialist rheumatological constructions of knowledge about musculoskeletal disorders. Using the concept of 'wear and tear' as our focus, we examine the way in which different understandings of these health problems are embedded in the question of the relationship between work or occupation and pain in the muscles and joints. Finally, we consider the role of general practice in relation to the different interests represented by these discourses, and how these are handled in the practical context of everyday management and therapy.

The shifting foundations of expertise

The relationship between experts, technical knowledge and society is central to sociological analysis (Bauman 1922, Bell 1973, Giddens 1990). Recently the sociologies of science, health and illness, and of the body, have coalesced around a problematisation of expert knowledge or discourse, and often a relativist stance in relation to the claims to knowledge made by both lay and professional experts. Disillusionment with biomedicine, it is argued, is one instance of a wider breakdown of public trust in the systems, experts and social phenomena associated with modernity. In the context of health and illness this dual process of 'secularization' and 'laicization' has made it necessary for individuals experiencing serious illness to elaborate their own frameworks of interpretation (Williams 1984).

Historians and historically-minded sociologists have documented how medicine's shift to a more 'scientistic' mode was characterised by an exclusion of the patient's view from discussions of the aetiology and nosology of disease. The very subjectivity of the patient's voice was thought to compromise that objectivity which has become the *sine qua non* of medical knowledge (Jewson 1976). With the development of new instruments for seeing inside the body, patients' accounts became less necessary to the practice of medicine (Pasveer 1989), and have been increasingly shaped by technological processes since the early twentieth century (Daly 1989). Although it has also been claimed that medicine developed an increasing interest in 'the patient's view' during this period, this interest was not in the patients' views 'in themselves', but only in the window they opened onto pathology (Armstrong 1984).

One prominent sociological response to this eclipse of the patient's view has been the development of a distinction between knowledge related to concepts of 'disease' – linked to biomedical knowledge – and the lay knowledge rooted in the experience of 'illness' (Eisenberg 1977, Williams and Wood 1986). Notwithstanding subsequent critiques entailed in this distinction (Hahn 1984), this dichotomisation of knowledge proved to some extent productive in generating an increasingly in-depth analysis of 'lay beliefs' in their social and biographical contexts.

While much of the early interest in lay perspectives took the form of an ethical critique of a *style* of medical superciliousness which disregarded patients' points of view, the critique has developed to a point where lay perspectives are now seen as forms of knowledge which are of 'equal worth' to expert knowledge (Stacey 1994). There are many different strands to this interest in forms of knowledge rooted in experience; and the willingness to see these forms as knowledge – rather than belief, prejudice or folklore – has emerged from a disillusionment with science and the growing scepticism amongst lay groups about the disinterestedness of science.

Employing a variety of different terms – 'lay knowledge' (Williams and Popay 1994, Popay and Williams 1996), 'people knowledge' (Stacey 1994), 'popular epidemiology' (Brown 1992), 'worker epidemiology' (Waterson 1993) 'citizen science' (Irwin 1996), and 'lay epidemiology' (Davison *et al.* 1991, Rogers *et al.* 1996) – social scientists working in different settings have begun to regard lay beliefs or the patient's 'point of view' not so much as interesting curiosities lying relativistically alongside or outside 'science', but more as forms of knowledge *sui generis*. Some of these examples involve lay people making use of knowledge produced by scientists and others involve an assertion of popular 'wisdom' against expert recommendations; some of them are embedded in political movements in which knowledge is collectively defined, while others are personal expressions of cultural stocks of knowledge framed by the exigencies of everyday life. However, all represent a refusal by lay people to see knowledge as something that should be defined by experts and limited to the controlled spaces of laboratories, seminars, and peer-reviewed academic journals.

Work by anthropologists and historians which seeks to illuminate the cultural and historical specificity of biomedical knowledge has sharpened the debate, and claims by biomedicine to neutral and universal knowledge are increasingly seen as revealing more about a particular social context than about the content of absolute truth (Gordon 1988). However, within the sociology of health and illness one consequence of the emphasis on lay knowledge has been the reinforcement of the distinction between 'lay' and expert', with a relative disregard for the ontology of expertise and the nature of expert claims to knowledge.

Sociology of scientific knowledge and the sociology of health and illness

A more clearly specified analysis of expert knowledge can be found in the sociology of scientific knowledge (SSK). The emphasis on the diversity of scientific and lay knowledge in this work provides a counterpoint to the understanding of both medical and lay knowledge as monolithic. The sources of technical knowledge, and the ways in which technical knowledge is integrated with other kinds of knowledge, are being explored in considerable detail – along with the different kinds of individuals and institutions which mediate such bodies of knowledge (Irwin and Wynne 1996). Although this field has its own argot, many of the issues with which it deals parallel concerns about the relationships between lay knowledge and professional knowledge within the sociology of health and illness (SHI).

Both SSK and SHI have been increasingly engaged in analyses of science and biomedicine which move beyond a critique of their social organisation to approaches which problematise and relativise scientific and medical knowledge. That is, in addition to criticising the élitism of scientists and the use to which science is put, the production of scientific and medical knowledge is now also regarded as a legitimate object of analysis by people outside science itself. However, while SHI has for the most part developed its critique of medical epistemology by way of an exploration of the particularity of lay knowledge, SSK has been much more caustic and direct in its attack on the claimed universality of the representational practices of science itself. Moreover, since the 1980s, SSK has been less inclined to treat 'science' as being of a piece in the way that SHI has tended to do in its critiques of biomedicine. In short, SSK has emphasised the heterogeneity of representational order in scientific knowledge, and the need to see 'knowledges' as socially negotiated and situated (Lynch and Woolgar 1990, Lambert and Rose 1996). Anthropological studies of medical practice resonate with this approach, and buttress the suggestion here that the practices of medicine operate in ways which are significantly different from the formal descriptions of biomedicine. (Hahn and Gaines 1985)

There are some indications that SHI and SSK are increasingly working with shared assumptions, though expressed in different ways. In addition to recent calls for the development of a 'strong programme' in medical sociology akin to that in SSK (Bartley 1990), the development of a more theoretically self-conscious concept of lay knowledge as opposed to 'lay beliefs' (see Gabe *et al* 1994, Popay and Williams 1996, Williams, S. and Calnan 1996) has led to a recognition of the need to deconstruct notions of expertise in modern societies. Empowered by the grand narratives of late modernity (Giddens 1990, Beck 1992), empirical work on the public understanding of science in SSK, and work on lay knowledge in SHI, has begun to draw attention to ' . . . the unacknowledged reflexive capability of lay people in articulating responses to scientific expertise' (Wynne 1996: 43). Such reflexively constituted lay knowledge is syncretic, developing from a process of 'bricolage' rather than formal training (Irwin *et al.* 1996: 55). In the context of the growing calls for methodological pluralism in public health and health services research (McKinlay 1993, Popay and Williams 1996) and the increasingly wide definitions of 'evidence-based medicine', such knowledge may in fact have greater ecological validity than that generated from the constrained context of the clinical trial.

However, in the context of experts' attempts to construe non-expert knowledge as 'discourses of ignorance' (Michael 1996), lay uncertainty about personal understanding of illness or local knowledge of pollution may lead lay people to devalue and de-legitimise their own knowledge regardless of how close in form and content to scientific knowledge their own understanding actually is (Lambert and Rose 1996). This reflexive devaluation of lay knowledge may be particularly evident in situations where lay people are directly engaged with the structures of science through the translation and legitimation purveyed to them by experts themselves. In situations such as that dealt with in SHI, lay perspectives are brought into direct contact with 'expertise' through relationships with *particular* experts. Moreover, in contrast to many of the examples from research in SSK, these experts – the physicians – are themselves relatively disempowered in relation to increasingly complex and diverse bodies of knowledge while continuing to carry the authority deriving in part from their relationship to that knowledge.

The struggle between the 'voices of medicine' and 'the voices of the life-world' (Mishler 1984) is not solely a theoretical one. The interpretive and critical relationship with medical knowledge developed by lay people acts to make scientific knowledge 'meaningful' by locating it in terms of personal experience, and is also critical of such knowledge in so far as it places it in the public domain as the subject for debate (Williams and Popay 1994). Knowledge, whether it is produced by lay or professional experts, is forged in social relationships.

Musculoskeletal disorders: exploring different dimensions of knowledge

In studies of lay knowledge of musculoskeletal disorders the concept of 'wear and tear' features as a key orientating notion. In the context of what has been said about the presence or absence of science within lay accounts it is an interesting idea because it is one that is used within *both* lay and biomedical representations of 'arthritis'. While lay ideas about musculoskeletal disorders often use 'wear and tear' as a concept to connect joint pain to their work, possibly influenced by social class (Elder 1973), clinicians often use the term to describe the appearance of erosions in the joints regardless of any external trauma. Before going on to develop a deeper analysis of some of the meanings and implications of 'wear and tear', we will outline some of the different contexts in which knowledge about joint pains has developed.

Specialist knowledge

Consideration of the history of 'rheumatological' activity suggests that medical practitioners' involvement in joint problems has revolved around rehabilitation medicine: exercises, manipulations and hydrotherapy characterised the treatment for these kinds of disorders (Kersley and Glyn 1991). In seeking to move 'beyond' its roots in the spa hospitals, rheumatology specialists engaged in a continuing and often fraught process of distinguishing themselves from other specialists, such as bone setters and osteopaths, those in physical and rehabilitation medicine, and more recently in orthopaedics.

The discovery of cortisone in the late 1940s played a significant part in the emergence of rheumatology as a more established specialism (Cantor 1993) but the early promise of this miraculous breakthrough was soured by an emerging awareness of its limitations. There have been few further breakthroughs, and the efficacy of more recent disease-modifying anti-rheumatic drugs has been markedly limited. Advances in this field are generally associated with developments in differentiating between, and in categorising, different forms of disease, and the technologies of surgical intervention and imaging.

More common disorders such as osteoarthritis and back troubles – associated with a major burden of disease and suffering – are relatively neglected by scientific and medical authors in the pages of contemporary rheumatological journals (Williams *et al.* 1992). Indeed, it may be that the very ordinariness of osteoarthritis, together with its association with ageing and mechanical stresses and strains, has militated against professional interest in it. The formal knowledge base of biomedicine contains two kinds of theories about osteoarthrosis and degenerative joint diseases: those which emphasise endogenous, possibly inevitable degeneration, and those which emphasise damage induced by various kinds of trauma (Holt 1973).

Given the competition between different medical specialties, it is perhaps not surprising that rheumatology should present itself as dealing with 'a number of clinically very challenging disorders including RA [rheumatoid arthritis], SLE [systemic lupus erythematosus], progressive systemic sclerosis, polymyositis, and Wagner's granulomatosis' (Williams 1994: 1097). The picture of a relatively beleaguered profession can be contrasted with the excitement conveyed in its professional journals about discoveries 'at the frontiers of knowledge', mostly relating to the rarer forms of disease such as those mentioned. The highly specialised body of knowledge within contemporary rheumatology itself draws on developments from immunology, biochemistry and genetics.

At the same time a vacuum of understanding exists in the area between high science and the development of strategies for disease management and patient care. Notwithstanding the development of criteria for the diagnosis of some specific disorders, the majority of disorders of the muscles and joints remain difficult to diagnose precisely and treat effectively. Rheumatologists themselves sometimes recognise a 'therapeutic impasse in osteoarthritis' despite increasing knowledge of pathophysiological events. However, the professional response to this recognition is to identify 'opportunities' to address this impasse through 'cross-disciplinary research in cartilage biochemistry, cell and molecular biology and biomechanics' (Herman and Vess 1994: 2000) rather than an engagement with the reality of lay experience and primary care.

Lay knowledge

The experience and understanding of joint pain is not merely common to large numbers of people, it is part of a stock-of-knowledge-at-hand which is embedded in everyday life. Jeremy Seabrook, for example, provides this graphic description of the life and circumstances of his grandfather in the context of a sociological essay on the rise and fall of the labour movement:

> My grandfather was a craftsman, a hand-sewer of shoes with awl and thread. During his early years as a skilled man he earned a living wage; but as his family grew and trade varied, his skill failed to provide enough to feed and shelter them. Later, his fingers became arthritic; he did more casual and unskilled work. He drank, and his family sank deeper into poverty (Seabrook 1978: 11).

'Arthritic' in this account, is clearly part of a complex chain of cause and effect connecting this man's body to his circumstances, and intimately related to the political economy of his life and times. Seabrook's account is an impressionistic one which shows us how musculoskeletal symptoms are part of a context and a trajectory. Although he does not go on to tell us what role, if any, the medical profession of the time might have played in diagnosing, explaining, and managing his grandfather's musculoskeletal problems, we are left with the understanding that they were probably unable to do very much, even supposing he were on a doctor's panel.

Within the sociology of health and illness musculoskeletal disorders have provided a focus for thinking about chronic illness. Studies have examined various components of daily life with disorders such as rheumatoid arthritis (Wiener 1975, Bury 1982, Locker 1983, Williams 1987), exploring the meaning of illness in terms of both its consequences and its significance (Bury 1991). In particular, research has explored the complexity of lay beliefs about the causes of arthritis, and interpreted the models people develop in terms of a narrative reconstruction of the relationship between body, self, and society (Williams 1984). Studies have also examined the intricate strategies people with arthritis develop for managing the many dimensions of the impact of chronic illness on their roles and relationships in both private life and the public sphere (Bury 1991, Locker 1983, Thornquist 1995, Newman *et al.* 1996).

Most of the work on musculoskeletal disorders has examined the experience of people with a particular diagnosis. However, while the existence of a diagnosis implies the important contribution that lay-professional interaction makes to the construction of both lay and professional knowledge, this is rarely drawn out. In the study we now go on to describe, we have taken as a starting point the existence of lay knowledge and then examined not only some of the sources people drew on, but also the relationship of such knowledge to decisions about seeking health care.

Research design

The research to which we refer is based on a study of musculoskeletal disorders in the community which combined epidemiological, clinical and sociological sources of data. A population survey of approximately 6000 adults registered with GPs established a sample of adults with self-reported pain in muscles or joints. Experience of such symptoms – not diagnosis or contact with specialist services – was the core criterion for inclusion in all subsequent stages of the study. Of these, 580 'positive' (that is, symptomatic) respondents were selected to attend a research clinic which enabled clinical data and diagnoses to be established.

From this group a sub-sample of 80 respondents was selected for in-depth interviews. The purposive selection of this 'qualitative' sample was based on an intention to seek more in-depth understanding of the use people make of their GPs and other practitioners; the sample included people who had not consulted anybody about their symptoms and those with very intensive patterns of using healthcare. It is therefore not representative in the statistical sense, including as it does more 'extreme' cases in terms of both need and utilisation than is likely to be the case with the larger sample. However, in terms of demographic composition, range of symptoms and co-morbidity the clinic and qualitative samples are comparable. (For further details of methodology and the qualitative subsample see Note at end of text.) Where respondents have been cited in this text their study numbers have been given in brackets, and some of the key characteristics are summarised in Table 1.

The interviews were undertaken using a topic guide which focused on knowledge and beliefs (in particular views about the cause(s), treatment and prognosis of symptoms or disorders) in relation to patterns of help-seeking in the broadest sense. We were interested in looking at how and why decisions were made to seek or not to seek certain kinds of help from within and outside formal primary care. The open structure of the interviews enabled respondents to frame their responses within their own priorities if they wished and to raise additional concerns.

In this chapter, the data from these respondents is counterpoised with consideration of clinical perspectives. For the latter we have drawn on two sources of data. The first is the professional literature about musculoskeletal disorders. The second is our fieldwork: dialogue with professionals occurring in the course of this research supplemented by recorded interviews with a rheumatology consultant and with three academic GPs. Although not constituting a systematic study, the fieldwork and interviews, in conjunction with the professional literature, do provide enough of the context of medical knowledge to allow us to begin to explore the relationship between lay and clinical knowledge.

The process of 'wear and tear': the body in labour

An association – negative or positive – between work and health or well-being is deeply embedded within the cultural life of industrial, de-industrialising and post-industrial societies. Many somatic and psychological

Table 1. *Details of respondents cited in text*

Study number	*Gender*	*Age*	*Site of pain*	*Work status*	*Occupation/former occupation*	*Health practitioners consulted for musculoskeletal symptoms*
286	Male	71	Neck pain	Retired	Engineering (manual) work	None.
372	Male	57	Multiple joint pains	'Retired' on grounds of disability	Painter/decorator	GP, hospital consultant, physiotherapist, herbalist.
538	Male	57	Neck pain	Retired	Plumbing/work for waterboard	GP, pharmacist, physiotherapist, herbalist.
656	Female	53	Multiple joint pains	'Unemployed' – not able to work due to sickness	Post woman	GP, hospital consultant, acupuncturist, physiotherapist
1461	Female	46	Shoulder pain	Self-employed	Child minder	GP.
3309	Female	46	Multiple joint pains	'Unemployed' – not able to work due to sickness	Factory work	GP, hospital consultant, pharmacist, physiotherapist.
4462	Female	54	Multiple joint pains	Working part time	Shop work	GP, hospital consultant

Note: All respondents cited were given a diagnosis of osteoarthritis or – in two cases – degenerative disc disease, at the research clinic.

disorders are thought to be the product, in some way, of the presence or absence of work. Musculoskeletal disorders include numerous symptoms associated in one way or another with particular kinds of occupational activity. Some of the earliest and most influential epidemiological studies of musculoskeletal disorders examined the experience of coal miners and identified work-related patterns of symptomatology (Miall *et al.* 1953). Moreover, in the patois of the coal miners themselves, terms such as 'beat knee, beat elbow and beat hand' were often used to describe the effect of coal mining upon the muscles and joints (Coombes 1944: 51). More recently, primarily in the context of litigation by employees in Australia, there has been much analysis and debate concerning the status and origin of repetitive strain injury (RSI) (Meekosha and Jacubowicz 1986, Arksey 1994).

One respondent (538) from our study, who associated the onset of his back and neck problems with trauma at work, developed his answer further when asked about 'anything else you think might have caused the arthritis'. He mentioned that he had experienced rheumatism as a child, and that he wondered whether this may have made him more susceptible, that 'people with red and fair hair (like himself) have a tendency to be arthritic', and finally returned to develop the theme of hazards at work as a cause. He talked about using a pneumatic drill frequently at work in his twenties, which 'shakes you from head to foot' and 'doesn't do the spine any good', and finally summed up the causes of his arthritis as 'wear and tear'.

The elaboration of the concept of wear and tear often drew respondents into more detailed reflection on how their bodies had been 'constrained' by their working lives:

> . . . it started from a back injury through some scaffolding collapsing . . . they assessed me for percentage of disabled and I got a green card so then I carried on doing the job I was doing but luckily I got a foreman's job then . . . so it really took a lot of pressure with regard to doing physical work . . . (372).

Another respondent, now in her forties, was employed by the Post Office for many years. She was compelled to retire on the grounds of ill-health after her entitlement to sick leave ran out: suffering from multiple joint pains and stiffness, this woman was no longer able to continue delivering mail. She identified having to move house and other stresses as being associated with 'bad phases' when she suffered more pain. However, at the centre of her explanatory framework were the constraints of her work:

> I've worked at the post office delivering mail and, like I said, I've always had twinges in me back if I'd sat down for a while, but it got worse . . . I've worked there thirteen years at the post office, and for about the last three years [. . .] it was getting worse, me back, and it was affecting me leg, I'd be walking, walking around posting the mail [. . .] but it was the

bending down, you know, to the letter boxes, most of them was at the bottom of the door and, of course, carrying the weight . . . (656).

Respondents often emphasised the same specific features of the workplace more than once, providing a 'thick' description of their past and present circumstances as an interpretive context for making sense of their symptoms. What also comes through is the sense of their having little room for manoeuvre in the deployment of their bodies in relation to the work. Their experience of embodiment can only be understood within this context. While sociologists may have neglected human embodiment in favour of the meaning of social action (Turner 1992), it is clear from these extracts that – in almost theological terms – the embodiment of 'suffering' only has meaning in relation to '. . . the deeds and decisions of life' (Macquarrie 1966: 276).

Many of the older respondents talked about the onset of pain associated with a heavy burden of manual work. Quite often they were able to evoke a sense of the sheer physical struggle of manual work in a harsh environment, or the qualities of the actual material with which they were working. Talking about the genesis of his symptoms, one man who had worked in a variety of manual jobs argued:

I was doing engineering work. On a heavy day it would consist of me shifting heavy metal castings which sometimes could weigh half a ton [. . .] and of course in the woodwork industry you get 8 x 4 sheets of timber which are . . . even though they are not so heavy as metal work . . . they are awkward to handle, and I think basically that's the only problem that I've had [. . .] lifting awkwardly (286).

Other retired respondents emphasised lives of work and hardship, and in this case several felt far more able to manage symptoms once they had stopped work. Many of this older generation had reached an accommodation with their former employers, where they had retired on grounds of ill-health after many years of work. For these respondents an adequate description of the relationship between body and social context requires more than a flat description of the job and the site of their pain. It calls out a complex representation of the nature of their occupation, the specific bodily movements their job required of them, their relationship to the materials of their trade, the involvement of other people, and the way in which all these things had an impact on how they felt.

Those still in employment (who worked mainly in the service industries and the public sector) reported pressures to 'keep going' under extremely difficult circumstances. The few self-employed people within the group were not exempt from these kinds of pressures. As one childminder put it:

. . . we weren't designed to push pushchairs more than for a couple of years at a time – keep doing it all the time [and] it's bound to have some effect (1461).

In addition to the detailed representation of the relationship between the body and the context of labour, there is a sense being conveyed of something wrong about the way we live; something unfree and unnatural. This sense of there being something not right about having to repeat the same body movements was common to many of the explanations of cause of joint pain, and more usually occurred in accounts of factory work, lifting patients in hospitals, or other kinds of employment involving lifting and pushing the same objects day in day out. This was combined in the accounts with an experience of not being able to stop, of having to go beyond one's limits.

It was not that our respondents protested or complained explicitly to us about the kinds of conditions of work in which they were obliged to operate, but there were occasions when such conditions sounded quite intolerable. Respondents spoke of working 80–100 hours a week, for example, or working all day within a very small space with only very limited scope for movement. In both temporal and spatial terms the essence of work was constraint, and it was clear that respondents felt so 'worn out' with trying to manage these obligatory pressures in the context of growing pain, that wear and tear in the joints was often seen as not surprising.

Lay responses to professional expertise

In a recent series of briefings on current knowledge and practice published in the *British Medical Journal* (Smith 1996) leading medical experts included as key factors implicated in the development of OA, 'mechanical stresses' and 'trauma to the joints', both of which fit incontestably into the concept of 'wear and tear'. The aetiology of degenerative musculoskeletal disorders is described in terms of a multifactorial model which juxtaposes such mechanical 'stressors' alongside others such as heredity and psychological stresses. At first glance, therefore, there appears to be substantial agreement between lay and clinical knowledge about what causes 'degenerative' musculoskeletal disorders.

The differences between lay and professional perspectives emerged piecemeal in respondents' accounts of knowledge in practice, though most of our respondents were reluctant to express disagreements with their doctors overtly. Many of the lay accounts were structured around a prolonged series of attempts to seek help for pain and discomfort which could be distressing and isolating. In keeping with the inevitability accorded to the 'wear and tear' of joints in lives of hardship, people generally saw arthritis as an illness with no cure – 'if you have arthritis, you'll have it until they send you out in a black box'. However, there was often a kind of triumph of hope over knowledge: while something was being done, even if that was only waiting for the results of tests, hope could be sustained.

It was in this context that access to specialist knowledge and technology were important, but this required difficult negotiations with general medical practitioners (GPs) who remain the first point of contact with the health care system for most users of the British NHS:

> I kept going back to the doctors . . . Anyway it got that way I went three weeks when I was in agony and I went to him and I said, 'well I want to see a specialist', and he said, 'I don't think there's anything wrong'. I said, 'well I do and if you're not going to send me to see one I want to pay to see one'. So he said, 'you can't do that', and I said, 'yes I can'. So he said, 'I'll write you a letter'. Two weeks later he hadn't written me a letter, right, so I got on the phone and I said I want a letter . . . (4462).

At the interface between primary and secondary care the general practitioner's 'cultural authority', or the power to have medical definitions of reality accepted, can translate almost seamlessly into the 'social power' to control actions relating to health care, paradigmatically in decisions about referral from primary to secondary care (Starr 1982, Elston 1991). In addition to being gatekeepers of the demand for care, GPs are the point at which lay knowledge comes into contact with the world of professional expertise. The GP acts as the lens through which the history and current state of expert knowledge is refracted to users of health services. In his or her relationship with patients the GP embodies the tension in the relationship between biological knowledge of pathology and sociological knowledge of person and context.

Because of the GP's relative proximity to the world of lay experiences and to limited specialist medical and bioscientific knowledge, it is here that the complex relationships between lay and expert knowledge are to be most clearly discerned. In contrast to the detailed contextualisation of the circumstances surrounding 'wear and tear' in lay accounts, within primary care the inevitability of joint pain linked to wear and tear was often described in terms of the inevitability and blamelessness of the ageing process.

Respondents found the idea of ageing leading to wear and tear and then to joint symptoms difficult to disagree with. However, in contrast to the detailed experiences of their own working lives, it often seemed an unsatisfactory explanation when applied to the particulars of a person's life. There was also a feeling that explanations in terms of ageing were a way of saying that nothing could have been different and that nothing can now be done. The association of musculoskeletal degeneration with ageing had left them with very little help, whatever their age actually happened to be. In the case of osteoarthritis and other non-specific, diffuse joint pain, the apparent consensus about the inevitability of natural 'wear and tear' may lead to an agreement about the limits of medicine which partly accounts for the high proportion of people with musculoskeletal disorders who do not seek help.

Opportunities for effective intervention are limited, and uncertainty about both aetiology and prognosis predominates. This viewpoint is not merely a sociological conceit. The absence of effective treatments for most musculoskeletal problems is central to many of the recommendations about medical management (Smith 1996). For those 'degenerative' osteoarthroses which constitute the major burden of illness, the emphasis, even in the orthodox literature where we might expect a strategic emphasis on what physicians can do, tends to be on the management of demand in the context of a gap between a limited capacity to intervene effectively and patient expectations of, or demands for, help.

In our discussions and interviews with GPs a recognition of the threat to the doctor/patient relationship was associated with their limited capacity to intervene therapeutically in these conditions. They indicated that referral for specialist opinion or treatment was sometimes undertaken in order to maintain trust in the relationship, rather than for narrowly clinical reasons. GPs recognised that the management of a patient's needs in the context of their relationship was complex. As one GP put it:

> I don't refer somebody with osteoarthritis to a rheumatologist unless there is a problem with my relationship with the patient and I can't get that patient to be reassured that this is a simple osteoarthritis.

Another GP recognised that there was a degree of collusion in this process:

> The GP all too often knows there isn't anything special going to come from the specialist but he can't think of anything else to do . . . there is definitely a minor conspiracy that goes on which is not shared with the patient. The GP on the whole does not say to the patient well you've asked me to refer you to a specialist and I think it's going to be a complete waste of time.

However, in these accounts there is a sense of the limited extent to which GPs grasp the complex knowledge-guiding interests which inform the bricolage of lay knowledge. In their accounts of seeking help people emphasised appointments with specialists at hospitals as pivotal. These appointments seemed to be significant not only because of their practical value (as a gateway to other services such as physiotherapy, for example), but also as a basis on which to interpret and make decisions. The uncertainty and the provisional nature of encounters with specialists makes them in some ways less alienating than the management of demand which much primary care represents. Specialist knowledge can provide patients with new possibilities for reinterpretation and action. 'Specialist', in the context of lay bricolage, does not mean only what is encountered in orthodox secondary and tertiary services. The interventions of some alternative therapists (such as osteopathy and chiropractors) were seen as drawing on a specialist knowledge base which could translate into action or intervention. Whether approaches were 'orthodox' or 'unorthodox' was of less interest than whether they were specific enough to have an (effective) intervention based on them. This valuing of specialist knowledge contrasted with a view of GPs as too often lacking the specific knowledge to intervene effectively in musculoskeletal problems.

It was against this background of the nature of lay needs for knowledge that 'tests' recurred like a litany in accounts by our respondents. Some people said they had only begun to have an understanding of what was wrong after seeing the X-ray, and several people who had not seen their X-rays expressed a wish to do so. A rheumatologist whom we interviewed acknowledged sometimes ordering X-rays 'because they [patients] feel they need something to feel that their problem has been taken seriously.' In some cases the doctor's interpretation of the X-rays had provided an image of what was wrong which was deeply congruent with their experience:

> If I could I'd work. I'm not soft and, yes, I am in pain. It shows up on the X-rays that I've got a disc out at the bottom of my spine. Now Mr M did say I was in so much pain because with your disc being at the bottom it's not strong enough to hold your body. It feels like my spine is going into my bottom. It's like I want chopping in half, my back is too heavy for me (3309).

One person had taken a photograph of an earlier X-ray and wished to do so of a more recent one after an operation on his knee, in order to compare before and afterwards. Far from being an unwarranted invasion of the self, X-ray technology is in some cases being appropriated by lay people to further their own needs for knowledge of their bodies, and can provide physiological evidence to connect to the sociological account of the origins of wear and tear.

Conclusion

Eliciting and interpreting models of lay beliefs about the causes of disease is now a well-established domain of sociological inquiry. In this chapter we have attempted to develop this area of work by focusing on the meaning of a particular category of aetiological and diagnostic knowledge, 'wear and tear', and examining the way in which it is connected to the domain of medical knowledge. In particular, we have shown how lay accounts about joint pain are grounded in a detailed interpretation of the nature of work; not only in terms of its constituent activities, but also the social processes of pressure and constraint which constitute the reality of labour. We have suggested that lay understanding of some musculoskeletal disorders is developed out of a response to the intolerable, particularly in relation to work situations. There is considerable ambivalence amongst GPs about their role in relation to legitimating (or withholding legitimation of) illnesses related to wider economic hardship and social deprivation. In some situations lay people may initiate or collude with the medicalisation of their distress in order to access the benefits and support they need to manage their symptoms in their everyday lives. An undue emphasis in the analysis of medicalisation on a unilateral imposition of meaning on patients/bodies will not help solve current dilemmas.

Whereas the 'multifactoriality' of 'expert' accounts involves the addition of more and more variables into putative causal models, lay explanations involve the interpretive elaboration of the meaning of one or more factors in the context of everyday lives. The 'multifactoriality' of lay accounts consists less of the proliferation and sub-division of factors and more of the interpretive exploration of the meaning of one or two ontologically central factors – in this case work – and their relationship to aspects of human embodiment. The relationship between 'work' and the experience of symptoms makes sense, both epistemologically – in terms of their knowledge of how things have affected them – and ontologically – in being part of the framework of their identity.

We have also shown that in practice the reification of 'lay' and 'professional expert' knowledge is not helpful. The bricolage of lay perspectives is an attempt to develop a body of knowledge that draws on a variety of sources in order to inform processes of interpretation and action. In this process, specialist bodies of knowledge and experts' sources of all kinds may be useful. Expert knowledge in this setting too is a curious mixture of homely moralising and arcane scientism which varies depending on the context in which it is being employed. The analysis of expert knowledge in this context is less clear-cut than it is in much of the work in SSK discussed earlier and such knowledge is far from being equally available to all practitioners dealing with musculoskeletal disorders.

GPs in particular are faced with the challenge of providing bridges between the uncertainties of clinical practice and biomedicine's claims to epistemological privilege. They are caught between a yearning for some kind of scientific home which is increasingly irrelevant to the challenges they face, and uncertainty about how to respond meaningfully and thoughtfully to the social context of their patients' experience. For physicians situated between a prevailing professional ethos which emphasises attending closely to patients' accounts, and the fundamental constructions of biomedical science, 'resting [as they do] on comprehending [internal, biological, pathological] material processes' (May *et al.* 1996: 200), the strains of a medical 'multifactorial' model may be particularly evident in practice.

In this context, the promotion of GPs as advocates of the patient's perspective can be seen as a strategy of re-professionalisation, fortuitously strengthened by current drives towards an NHS 'led' by primary care. Although health services research has begun to delineate the significance of the GP's enhanced role in providing and shaping services, the implications of the location of the GP at a point between the specialist and the lay person

(Horobin 1983), as a mediator, facilitator and translator of knowledge, are perhaps less developed. The GP is potentially not only a primary care provider and a purchaser of specialist services, but a nodal point in a network of knowledge that includes lay knowledge itself, and various forms of complementary medicine, in addition to a range of orthodox services. In the context of current debates about skill mix and workforce planning in the NHS (Sibbald 1996), GPs' qualifications to undertake this role successfully are not evidently more appropriate than those of other health care workers.

The research reported here suggests that there are some major problems in the task of simultaneously conveying information and managing uncertainty. Many respondents in this study testify to occasions when such communication was unhelpful or unacceptable. These 'meetings between experts' (Tuckett *et al.* 1985) can often lead to a devaluation of lay knowledge, however meagre the professional's own knowledge may be. The development of lay discourses about these kinds of conditions can be seen as reconstructions of elements of this very ambiguous knowledge in the context of other forms of knowledge about identities, roles, relationships and social structures.

Acknowledgements

The research to which this chapter refers, funded by the Department of Health and West Pennine Health Authority, is part of the Population Health and Demand for Care programme at the National Primary Care Research and Development Centre, Universities of Manchester, Salford and York. The research is being undertaken jointly by members of the National Centre team at the Public Health Research and Resource Centre, University of Salford, and the ARC Epidemiology Research Unit, University of Manchester. We should like to thank Stephen Arcari and Bob Skelton for their support with this work, and the referees and editor of this monograph for the challenges posed by their expertise!

Note

Survey methodology and response
A postal screening questionnaire asking about musculoskeletal symptoms and disability was sent to an age- and sex-stratified sample of 5980 adults over 16 from the registers of three general practices in one health authority area. There was a response rate of 78 per cent to the survey. Of those who reported having 'aches and pains in muscles and joints' for longer than seven days within the past month (in response to a specific question on the questionnaire), a sample for each anatomical region was invited to attend a research clinic at the GP's surgery for assessment by a metrologist (a nurse trained in the measurement of joint pain) and consultant rheumatologist. The research clinic sub-sample also provided the sampling frame for the selection of respondents to be interviewed in depth.

Qualitative subsample
The subsample included men and women with a range of symptoms, periods of onset (from less than one year to over 20), experiences, circumstances and diagnoses. The majority were given diagnoses of osteoarthritis; other diagnoses included inflammatory diseases, mechanical back pain and fibromyalgia.

Although 33 were 'retired', this included those defining themselves as retired who had obtained 'early retirement' on health grounds; 17 worked full time; 14 were not working because of poor health or disability. Others worked part time, in the home, or were students. Occupations (and former occupations) ranged from heavy industry through factory work, to work in the service sector. The majority worked in skilled or semi-skilled manual occupations, or had done so in the past. Others had been employed in a variety of clerical, administrative, professional, and other non-manual public sector occupations. Of those who had retired a high proportion had been employed as manual workers in the factories and cotton mills which used to dominate the area.

(Because the importance of respondents' relationship with work was not an *a priori* focus, details of occupations for those in the larger samples were not sought. The importance of this relationship in the context of these kinds of symptoms emerged from an analysis of the qualitative data and will inform the development of subsequent research.)

Further details about the project are available in a report form.

References

Arksey, H. (1994) Expert and lay participation in the construction of medical knowledge, *Sociology of Health and Illness*, 16, 448–69.

Armstrong, D. (1984) The patient's view, *Social Science and Medicine*, 18, 737–44.

Bartley, M. (1990) Do we need a strong programme in medical sociology? *Sociology of Health and Illness*, 12, 371.

Bauman, Z. (1992) *Intimations of Postmodernity*. London: Routledge.

Beck, U. (1992) *Risk Society*. London: Sage.

Bell, D. (1973) *The Coming of Post-Industrial Society*. New York: Basic Books.

Brown, P. (1992) Popular epidemiology and toxic waste contamination, *Journal of Health and Social Behaviour*, 33, 267–81.

Bury, M. (1982) Chronic illness as biographical disruption, *Sociology of Health and Illness*, 4, 167–82.

Bury, M. (1991) The sociology of chronic illness: a review of research and prospects, *Sociology of Health and Illness*, 13, 451–68.

Busby, H. and Williams, G. (forthcoming) *Musculoskeletal Disorders in the Community: Explaining Health Needs and Patterns of Care*. Project report. National Primary Care Research and Development Centre.

Calnan, M. and Gabe, J. (1991) Recent developments in general practice. In Gabe, J., Calnan, M. and Bury, M. (eds) *The Sociology of the Health Service*. London: Routledge.

Cantor, D. (1993) Cortisone and the politics of empire: imperialism and British medicine 1918–1955, *Bulletin of the History of Medicine*, 67, 463–93.

Comaroff, J. (1985) *Body of Power, Spirit of Resistance: the Culture and History of a South African People*. Chicago: University of Chicago Press.

Comaroff, J. and Maguire, P. (1981) Ambiguity and the search for meaning: childhood leukaemia in the modern clinical context, *Social Science and Medicine*, 15B, 115–23.

Coombes, B.L. (1944) *Those Clouded Hills*. London: Cobbett Publishing Co. Ltd.

Daly, J. (1989) Innocent murmurs: echocardiography and the diagnosis of cardiac abnormality, *Sociology of Health and Illness*, 11, 99–116.

Davison, C., Davey Smith, G. and Frankel, S. (1991) Lay epidemiology and the prevention paradox: the implications of coronary candidacy for health education, *Sociology of Health and Illness*, 13, 1–20.

Eisenberg, L. (1977) Disease and illness: distinctions between professional and popular ideas of illness, *Culture, Medicine and Psychiatry*, 1, 9–23.

Elder, R.G. (1973) Social class and lay explanations for the etiology of arthritis, *Journal of Health and Social Behaviour*, 14, 28–38.

Elston, M.A. (1991) The politics of professional power: medicine in a changing health service. In Gabe, J., Bury, M. (eds) *The Sociology of the Health Service*. London: Routledge.

Gabe, J., Kelleher, D. and Williams, G. (1994) *Challenging Medicine*. London: Routledge.

Giddens, A. (1990) *The Consequences of Modernity*. Oxford: Polity Press.

Gordon, D. (1988) Tenacious assumptions in western medicine. In Lock, M. and Gordon, D. (eds) *Biomedicine Examined*. Kluwer: Academic Publishers.

Hahn, R. (1984) Biomedical practice and anthropological theory, *Annual Review of Anthropology*, 12, 305–33.

Hahn, R. and Gaines, A.D. (eds) (1985) *Physicians of Western Medicine: Anthropological Approaches to Theory and Practice*. Holland: D. Redel.

Herman, J. and Vess, E. (1994) Therapeutic impasse in osteoarthritis, *British Journal of Rheumatology*, 33, 12, 1098–2000.

Holt, L. (1973) Articular cartilage. In Gilliland, I. and Peden, M. (eds) *The Scientific Basis of Medicine Annual Reviews: 1973*. London: Athlone Press/Postgraduate Medical Association.

Horobin, G. (1983) Professional mystery: the maintenance of charisma in general medical practice. In Dingwall, R. and Lewis, P. (eds) *The Sociology of the Professions*. London: Macmillan.

Irwin, A. (1996) *Citizen Science*. London: Routledge.

Irwin, A. and Wynne, B. (eds) (1996) *Misunderstanding Science? The Public Reconstruction of Science and Technology*. Cambridge: Cambridge University Press.

Jewson, N. (1976) The disappearance of the sick man from medical cosmology: 1770–1870, *Sociology*, 10, 224–44.

Kersley, G. and Glyn, J. (1991) *A Concise International History of Rheumatology and Rehabilitation: Friends and Foes*. London: Royal Society of Medicine.

Lambert, H. and Rose, H. (1996) Disembodied knowledge? Making sense of medical science. In Irwin, A. and Wynne, B. (eds) *Misunderstanding Science? The Public Reconstruction of Science and Technology*. Cambridge: Cambridge University Press.

Locker, D. (1983) *Disability and Disadvantage: the Consequences of Chronic Illness*. London: Tavistock.

Lynch, M. and Woolgar, S. (eds) (1990) *Representation in Scientific Practice*. Cambridge, MA: MIT Press.

Macquarrie, J. (1966) *Principles of Christian Theology*. London: SCM Press.

McKinlay, J. (1993) The promotion of health through planned socio-political change: challenges for research and policy, *Social Science and Medicine*, 36, 109.

May, C., Dowrick, C. and Richardson, M. (1996) The confidential patient: the social construction of therapeutic relationships in general medical practice, *Sociological Review*, 44, 187–203.

Meekosha, H. and Jacubowicz, A. (1986) Women suffering RSI: the hidden relations of gender, the labour process, and medicine, *The Journal of Occupational Health and Safety – Australia and New Zealand*, 2, 390–410.

Miall, W.E., Caplan, A., Kilpatrick, G.S. and Cochrane, A.L. *et al.* (1953) An epidemiological study of rheumatoid arthritis associated with characteristic chest X-ray appearances in coal workers, *British Medical Journal*, (iv), 1231–6.

Michael, M. (1996) Ignoring science: discourses of ignorance in the public understanding of science. In Irwin, A. and Wynne, B. (eds) *Misunderstanding Science: The Public Reconstruction of Science and Technology*. Cambridge: Cambridge University Press.

Mishler, E. (1984) *The Discourse of Medicine: Dialectics of Medical Interviews*. New Jersey: Ablex.

Newman, S., Fitzpatrick, R., Revenson, T., Skevington, S. and Williams, G. (1996) *Understanding Rheumatoid Arthritis*. London: Routledge.

Pasveer, B. (1989) Knowledge of shadows: the introduction of X-ray images in medicine, *Sociology of Health and Illness*, 11, 4, 360–81.

Popay, J. and Williams, G.H. (1996) Public health research and lay knowledge, *Social Science and Medicine*, 42, 5, 759–68.

Rogers, A., Pilgrim, D. and Latham, M. (1996) *Understanding and Promoting Mental Health: a Study of Familial Views and Conduct in their Social Context*. Health Education Authority.

Seabrook, J. (1978) *What Went Wrong? Working People and the Ideals of the Labour Movement*. London: Gollancz.

Sibbald, B. (1996) Skill mix and professional roles in primary care. In National Primary Care Research and Development Centre, *What is the Future for a Primary Health Care NHS?* Oxford: Radcliffe Medical Press.

Smith, M. (ed) (1996) *ABC of Rheumatology*. London: BMJ Publishing Group.

Stacey, M. (1994) Lay knowledge: a personal view. In Popay, J. and Williams, G.H. (eds) *Researching the People's Health*. London: Routledge.

Starr, P. (1982) *The Social Transformation of American Medicine*. New York: Basic Books.

Thornquist, E. (1995) Musculoskeletal suffering: diagnosis and a variant view, *Sociology of Health and Illness*, 17, 166–92.

Tuckett, D., Boulton, M., Olsen, C. and Williams, A. (1985) *Meetings between Experts: an Approach to Sharing Ideas in Medical Consultations*. London: Tavistock.

Turner, B. (1992) *Regulating Bodies: Essays in Medical Sociology*. London: Routledge.

Watterson, A. (1993) Occupational health in the UK gas industry: a study of employer, medical and worker knowledge and action on occupational health in the late nineteenth and early twentieth century. In Platt, S., Thomas, H., Scott, S. and Williams, G. *Locating Health: Sociological and Historical Explorations*. Aldershot: Avebury.

Wiener, C. (1975) The burden of rheumatoid arthritis: tolerating the uncertainty, *Social Science and Medicine*, 9, 97–104.

Williams, G.H. (1984) The genesis of chronic illness: narrative reconstruction, *Sociology of Health and Illness*, 6, 175–200.

Williams, G.H. (1987) Disablement and the social context of daily activity, *International Disability Studies*, 9, 97–102.

Williams, G.H. and Wood, P.H.N. (1986) Common-sense beliefs about illness: a mediating role for the doctor, *The Lancet*, II, 1435–7.

Williams, G.H., Rigby, A.S. and Papageorgiou, A.C. (1992) Back to front? Examining research priorities in rheumatology, *British Journal of Rheumatology*, 31, 193–6.

Williams, G.H. and Popay, J. (1994) Lay knowledge and the privilege of experience. In Gabe, J., Kelleher, D. and Williams, G.H. (eds) *Challenging Medicine*. London: Routledge.

Williams, G.H., Fitzpatrick, R., MacGregor, A. and Rigby, A.S. (1996) Rheumatoid arthritis. In Davey, B. and Seale, C. (eds) *Experiencing and Explaining Disease*. Buckingham: Open University Press and The Open University.

Williams, R.G. (1994) Autoantibodies and rheumatology. *British Journal of Rheumatology* (Editorial), 33, 1097–8.

Williams, S. and Calnan, M. (1996) *Modern Medicine: Lay Perspectives*. London: UCL Press.

Wynne, B. (1996) Misunderstood misunderstandings: social identities and public uptake of science. In Irwin, A. and Wynne, B. (eds) *Misunderstanding Science? The Public Reconstruction of Science and Technology*. Cambridge: Cambridge University Press.

4. The rhetoric of prediction and chance in the research to clone a disease gene

Paul Atkinson, Claire Batchelor and Evelyn Parsons

Introduction

The gene responsible for myotonic dystrophy was located in 1991 by groups of scientists based in the UK (Cardiff and London), North America (Massachusetts Institute of Technology; Lawrence Livermore National Laboratory, California; Baylor College of Medicine, Texas; Ontario, Canada); and the Netherlands (Nijmegen) (Harley *et al.* 1992, Buxton *et al.* 1992, Aslandis *et al.* 1992, Brook *et al.* 1992, Fu *et al.* 1992). This significant scientific achievement allows patients whose family history indicates that they are at risk to be tested presymptomatically for Myotonic Dystrophy. The breakthrough relating to Myotonic Dystrophy is one among many genetic discoveries with direct relevance to clinical practice. Those scientific innovations have far-reaching implications for contemporary and future biomedical work. Indeed, there has been an explosive growth of molecular biology since the 1970s with the development of recombinant DNA technology, by which means scientists can manipulate DNA. This technology allows scientists progressively to narrow their search of the genome onto specific areas of DNA in their search for a disease gene. On the basis of this knowledge the scientists involved in the research on Myotonic Dystrophy claimed to have been able to predict that the Myotonic Dystrophy gene would be located and sequenced. In this chapter we report our research with members of one of the research groups responsible for the discovery of the gene. We explore how the scientists constructed their accounts of their work and of the 'discovery' in terms of scientific prediction, skill and chance.

While disease genes are being identified in increasing numbers, the task of searching for a given disease gene is immense (Connor 1993). Although recombinant DNA technology allows scientists to home in on a particular and increasingly narrow area of DNA, the search for a disease gene can take years. Some of the scientists we interviewed expressed the belief that, despite the progressive and incremental nature of the research, and the certainty that a disease gene existed, it was impossible to predict exactly when the Myotonic Dystrophy gene would be identified. Chance was described as having determined the pace of the research, especially in the final stages of the discovery. The scientists were also unable to predict with any degree of certainty the form that the genetic mutation would be found to take. Although the research was defined in terms of 'normal' science, the precise outcome was unforeseen and unexpected. Equally, despite the large-scale collaborative nature of the research, the rhetoric of chance was used to attribute success to particular groups and individuals.

Any scientific 'discovery' is not a unitary event in the natural world. What counts as a discovery is a matter of definition and interpretation – often retrospective – within one or more scientific networks (Brannigan 1981). Discoveries are constituted through series of socially organised actions and representations: publication, conference presentations, narrative accounts, discovery claims, priority disputes and so on. Discoveries and the significance attributed to them change over time. Narratives of discovery are occasioned, in that they vary with social context and audience. In this chapter we develop in greater detail one aspect of our work with genetic scientists, touched on elsewhere in a synoptic account of the research (Batchelor *et al.* 1996).

The sociology of scientific knowledge (SSK) implies a sceptical stance towards scientific discovery in general. This reflects a principled rejection of standard non-sociological accounts of scientific work, especially uncritical accounts of the history of scientific thought. A naive perspective on scientific knowledge portrays it in terms of a progression of discoveries, each successively reflecting a closer approximation to a true picture of the natural world. Popular accounts of scientific and medical discoveries, for instance, are likely to promote such imagery. Such emphasis on discovery thus reproduces a view of science that marginalises the course of controversies within and between networks of scientists, treats as irrelevant the processes of craft work and laboratory practice, and ignores the acts of interpretation and representation that constitute the phenomena of natural science. In contrast to the popular view, sociological perspectives stress that scientific discoveries do not establish themselves unequivocally: once a scientific discovery claim has been constructed it must be recognised by others and thus be accorded legitimacy by the relevant scientific community (Brannigan 1981). Discovery claims are important social phenomena within scientific networks. The discovery claim may gain considerable attention well beyond the scientific community. It is also a turning-point in many scientists' own career trajectories, and the period surrounding the discovery may be a key juncture for the research group (Batchelor *et al.* 1996). For all the epistemological scepticism of SSK, discovery claims in the natural and medical sciences remain important topics for sociological investigation.

Medical genetics offers productive sites for the exploration of medical and scientific knowledge production. It is one domain of medical work in which laboratory science and clinical practice are in especially close proximity. The 'clinical gaze' is increasingly supplemented or even displaced by a 'genetic gaze', through which an increasing number of disorders are attributed, at least in part, to genetically-determined dispositions. The field is changing rapidly, and scientific discoveries – such as the one described here – contribute directly to the geneticisation of medical science and clinical practice (Lippman 1991). Genetics is changing the way in which health and disease are conceptualised, and contributes to an increasingly pervasive discourse of 'risk' (Parsons and Atkinson 1992, 1993). Genetic science and 'the new genetics' pose increasingly complex problems of social and interpersonal decision-making (Marteau and Richards 1996). The coincidence of scientific and clinical work in the construction of genetic knowledge provides a major opportunity for the sociology of medical knowledge. The sociological exploration of medical knowledge in general represents one of the most significant developments for the contemporary sociology of medicine, health and illness (*e.g.* Atkinson 1995, Casper and Berg 1995, Clarke and Fujimura 1992, Fujimura 1987, 1988, 1992).

Much of our data (described in more detail below) consist of genetic scientists' retrospective accounts of the Myotonic Dystrophy gene discovery. We treat them as narrative accounts. We do not assume that they give us direct access to the scientists' personal experiences and feelings. In contrast, we do assume that they are shaped by repertoires of narrative and rhetorical devices, constructing accounts that in turn reflect shared cultural assumptions concerning scientific work and scientific discovery. This analytic stance owes much to the pioneering work of Gilbert and Mulkay (1982). Theirs remains one of the most thorough and sustained analyses of scientists' accounts of scientific innovation. Methodologically, it remains a model of how to treat informants' accounts of scientific (and other) work. In essence, Gilbert and Mulkay demonstrate how natural scientists use particular kinds of accounting device in their production of scientific accounts. These accounting devices are not consistent: indeed, they can appear to be mutually contradictory. They do not necessarily conform to established versions of scientific method and scientific research. Gilbert and Mulkay suggest that scientists' accounts of discovery are couched in terms of two different registers. Using accounting devices such as a 'truth will out' rhetoric, they account for discoveries in terms of conventional scientific inquiry, and the inevitable revelation of natural phenomena. These, however, alternate with a second register that stresses contingent, personal phenomena, using narrative devices that foreground scientists' personal characteristics, and unpredictable elements of chance. In our own examination of genetic scientists' narratives of their own success we explore how they too draw on narrative registers of predictive scientific inquiry and of chance in accounting for the discovery. In doing so we show how that discovery – from one perspective a predictable outcome of research within a normal-science paradigm (Kuhn 1970, Barnes 1982) – may also be a topic for considerable interpretative work by the scientists themselves, drawing on a variety of narrative devices.

To summarise our analytic perspective: for the constructivist sociology of knowledge, scientific discovery should be seen as a process of interpretation and representation rather than an event (Barnes 1982: 43). However, as Barnes argues, some discoveries are regarded as more predictable than others, confirming rather than overturning or changing theories. These anticipated discoveries therefore do not involve the process of reinterpretation required in cases of revolutionary science: they are consistent with scientists' expectations about the physical world, constituting work in normal science (Kuhn 1962). In the case of the Myotonic Dystrophy gene, the body of existing knowledge relating to the human genome was used in order to make an empirical prediction that a genetic mutation existed which caused the disease: the identification of the mutation was compatible with the scientists' expectations concerning human genetics. The genetic scientists' accounts identified the discovery as a predictable outcome of skilled and careful research. Embedded within the scientists' accounts of predictability, however, are rhetorical or narrative elements that stress chance and contingent factors.

The identification of the Myotonic Dystrophy gene cannot be viewed as an event occurring at any one moment in time. As Barnes (1982) argues, intrinsic to the very concept of discovery is validation, which cannot be seen as a single event. He argues that if a discovery is to be viewed as a unitary event at all, then it must be a psychological event, a perception that one is in possession of something. He even proposes that the term 'discovery' should be abandoned, although he recognises that the concept is useful for the scientific community and allows individual researchers to receive the status accorded to discoverers. The identification of the gene should not, therefore, be treated as a single moment, but as a process of socially organised cognition. It should also be viewed in terms of the socially validated results rather than the specific moment when any one scientist believed that he or she had discovered the gene.

The research: background and conduct

Ours was a small-scale study of the science and scientists involved in the isolation of the Myotonic Dystrophy gene. We were granted access to the Institute of Medical Genetics, at the University of Wales College of Medicine, Cardiff. Our fieldwork was conducted between October 1992 and May 1993, during the period immediately following the announcement of the Myotonic discovery. This qualitative inquiry was based on detailed interviews with members of the research group at the Institute of Medical Genetics at the University of Wales College of Medicine. We were also granted interviews with members of the research group at Charing Cross Medical School, London. In addition to those scientists' retrospective narratives of the discovery, we also collected their prospective accounts of research plans and the direction of the research group.

When discussing something as specific as a particular scientific claim, which is itself in the public domain through publications and other means, it is neither possible nor desirable to attempt to disguise totally the research site(s) and the actors involved. To do so would rob the sociological account of all specificity, remove a great deal of pertinent detail, and render the resulting account unreadably bland. We have, however, taken the precaution of trying to avoid the precise attribution of selected quotes to specific individuals. In presenting our findings and in attributing accounts, we have attempted to blur individual identities. Inessential details may have been changed somewhat with that end in mind. (It is for that reason that the data extracts are not indexed to specific individuals.)

The interviews were fully transcribed, and coded by means of the ETHNOGRAPH software for computer-aided qualitative data analysis (Seidel *et al.* 1988). They were then subjected to thematic analysis. In addition to conducting the interviews, the researchers also attended clinical and research meetings at the Institute of Medical Genetics, Cardiff. Further, documentary evidence from scientific and medical journals was obtained on the research in progress at a global level. This account is necessarily partial, as the overseas members of the collaborative network of scientists could not be interviewed. All were approached by letter and invited to contribute their views, and some did so. Since this paper concentrates on the narrative accounts we gathered, primarily from the Cardiff and London groups, the voices of the overseas collaborators are not available for exploration here.

It is necessary to locate the scientists' accounts against the backdrop of genetic research. One cannot make sense of the scientists' narratives without some background knowledge. Indeed, it is characteristic of research in this area that the social scientist must become at least acquainted with the technical knowledge that is the experts' stock-in-trade. The sociologist cannot simply maintain the stance of ignorant outsider without doing violence to the local knowledge and practical reasoning of the social actors involved. In the section that follows we do not repeatedly state that such 'background' information is itself a series of social products, and represents the socially agreed versions of knowledge current among a network of scientists, or inscriptions in journals and textbooks. To do so would prevent us ever reaching the substance of our analysis. Our shorthand presentation here does not, however, imply that we accept uncritically the following 'facts', or that we reserve constructivist analysis for just some aspects of genetic work, while privileging others.

Research into Myotonic Dystrophy

Myotonic Dystrophy is the most common form of adult muscular dystrophy (Harper 1989). It is an autosomal dominant disease with an incidence of

1:7,500, characterised by muscle wasting and weakness (Harley *et al.* 1992). The disease usually manifests itself earlier in the parent than in the child, and becomes progressively more severe throughout the transmission from one generation to the next. This phenomenon is known as anticipation.

The search for the Myotonic Dystrophy gene had been in progress in the UK and overseas since the early 1980s. Collaborative relations had developed between six of the research groups involved in the search. Although each research group adopted a slightly different approach in their search for the gene, based on their resources and skills, at each stage recombinant DNA technology was used to manipulate, clone, cut and splice the DNA. In the location of any disease gene, the particular chromosome on which the disease locus, or site, lies must first be identified. The research then increasingly focuses on smaller and smaller areas of the chromosome, until the actual locus is confirmed. The search, however, is not entirely straightforward since genes are not inherited in exactly the same form between generations: a process of recombination or exchange occurs between maternal and paternal chromosomes.

Using the molecular biological techniques, the research into Myotonic Dystrophy progressed in an incremental manner. In 1982 it was established that the Myotonic Dystrophy gene was on chromosome 19 and by 1987 a Nijmegen scientist had located the gene on the long arm of that chromosome. By 1989 scientists in Ontario and Cardiff, in the process of isolating DNA probes[1] which were linked to the Myotonic Dystrophy locus, had isolated markers which were very close to the Myotonic Dystrophy gene. Over the following years there was an increasing refinement of the location of markers which allowed the scientists progressively to home in on the gene. By 1991 Cardiff scientists had identified new polymorphic markers which were linked to the Myotonic Dystrophy locus. This was viewed as an important find, for the polymorphic marker was found to be inherited with Myotonic Dystrophy more frequently than would occur by chance. This work also indicated that most individuals affected with Myotonic Dystrophy in French-Canada and the UK were descended from a single ancestor. The Cardiff scientists thus concluded that it was possible that Myotonic Dystrophy was introduced into Canada over 300 years ago by one of the first founders. Throughout 1991 new markers were identified by means of screening DNA libraries which mapped to chromosome 19. Some of these were close to the Myotonic Dystrophy locus and two markers which flanked the Myotonic Dystrophy locus narrowed down the search even further.

The gene responsible for Myotonic Dystrophy was located in late 1991 and reported in the journals *Nature* (Harley *et al.* 1992, Buxton *et al.* 1992, Aslandis *et al.* 1992); *Cell* (Brook *et al.* 1992) and *Science* (Fu *et al.* 1992, Manadevan *et al.* 1992). The specific molecular defect is described as an unstable sequence of DNA which is located within the Myotonic Dystrophy gene. In unaffected individuals the nucleotides CTG are repeated no more than 30 times, while in affected individuals this can increase up to 2000 repeats. It is believed that the frequency of the triplet repeats is related to the severity of the disease and that this establishes a biological basis for the phenomenon of anticipation.

Discovery and confirmation

A PhD student working in the Charing Cross laboratory (in London) performed the experiment which first revealed the nature of the mutation. At the same time a Cardiff scientist was conducting a different experiment which revealed the identical mutation. Once the Myotonic Dystrophy gene had been identified it could not be immediately and unequivocally accepted by the scientific community. Scientists involved in the research had to judge the extent to which the phenomenon had indeed been located, and whether the predicted observation had been realised with adequate precision. It was therefore necessary to confirm the experimental results by replication.

Collins (1992) maintains that replication is '. . . the scientifically institutionalised counterpart of the stability of perception' (1992: 19). Notwithstanding its central place in the canon of scientific orthodoxy, the work of confirming experimental results is far from straightforward in practice. In his study of controversial scientific knowledge claims, Collins argues that while the idea of replication may appear to be straightforward, the practicalities involved in repeating an experiment or observation are problematic. In general terms, replicability is a complex process and scientists rarely bother to repeat others' findings unless they are particularly sceptical about them. Mere confirmation of others' results bestows few rewards on those who undertake the task. The scientific rewards – funding, career advancement and esteem – go to those who are the first to observe a new phenomenon or produce a new result. Those who repeat an observation or experiment gain little, if any, status themselves, but merely confirm the status of the original scientist(s). Furthermore, when replication is attempted, then what *counts* as a replication can be highly contentious and problematic. Failure to replicate is not necessarily interpreted as the falsification of the original experiment. It may be attributed to any number of factors, such as failure to reproduce the original experimental conditions with sufficient precision, through the application of inappropriate equipment, faulty technique and so on. Even when the outcomes of replication are regarded as predictable, and are seen as confirmatory in nature, the practical realisation of a replication may prove less straightforward than practitioners themselves may expect (Collins 1992). The general reluctance to undertake replications does not necessarily apply, however, in the critical phases of rapid change, or in the period immediately surrounding a new discovery claim.

There is an urgent need to test out the claims of other research groups on one's own data, or using one's own techniques. Research groups need to test whether new results can be incorporated into their own research designs, or can be confirmed using their own local techniques. In such cases, the process of replication does not compete with the desire to establish original research, or to establish priority claims. On the contrary, it becomes an urgent part of the race for significant new results. This was the case for the research groups involved in the Myotonic Dystrophy discovery.

In principle, the scientists working at Cardiff and elsewhere on Myotonic Dystrophy had little or no doubt that the experimental results would be replicated and confirmed: they believed that they were dealing with a 'normal science' problem, with a high degree of predictability. In the Myotonic Dystrophy research the independent scientific criterion to assess the quality of the final experiment was the cloning of the Myotonic Dystrophy gene and the identification of the mutation. However, while the presence of a genetic mutation had been predicted and was therefore an objective criterion for the success of the experiment, the exact form it would take was by no means predictable. The tensions between predictability and unpredictability in this discovery are expressed in the scientists' accounts explored throughout the remainder of our discussion. The experiments undertaken by the London and Cardiff scientists which resulted in unusual blots on the autorad,[2] were considered by members of the scientific community to be competently executed and therefore of value because they had detected the predicted mutation.

Once the London and Cardiff groups had confirmed their own results, scientific papers documenting the research were written. Because of the collaborative nature of the research the head of the London laboratory sent a draft of the paper detailing his group's findings to the Lawrence Livermore Laboratory in California for comments. The head of that laboratory then informed the Ontario and Nijmegen groups of the London findings. While he did not communicate the entire contents of the paper, he did pass on sufficient information to allow the Canadian group to reproduce the result within two days. The ability of the other groups to replicate the UK scientists' observations was due in part to the nature of the collaborative relationship whereby physical resources – such as probes – had been shared between groups on previous occasions. It was not necessary for the overseas groups to have precise details of the London experiment in order to validate the original results for they were already in possession of most of the data and resources needed to replicate the findings. As with the UK groups their research had also narrowed to the small area of DNA on chromosome 19. Because of the variability in scientists' approaches to the problem the overseas groups did not attempt precisely to replicate the experiments performed by the London group; rather, they sought to achieve the same result by their own local means. Thus, for example, members of the group based in the

Netherlands were able to identify the mutation using a mouse-brain cDNA.[3] They could therefore submit their findings to *Nature* for publication in tandem with the UK papers and so reap similar rewards.

Scientists' accounts of the research

Once a new scientific finding has been accepted and thus crystallised in the conceptual order, the negotiations and unscientific influences which occur before scientific knowledge emerged, are submerged (Collins 1992). Retrospective accounting for the production of the new knowledge concentrates on the scientific method whereby knowledge is produced by means of objective experimentation and observation to the exclusion of social contingencies. We found that in their retrospective accounts of the discovery the scientists often referred to contingent factors. In their study of scientific discourse, Gilbert and Mulkay (1982) identify two repertoires which scientists use in accounting for their actions and beliefs – empiricist and contingent. By means of the empiricist repertoire scientists are portrayed as impersonal observers and manipulators of the natural world; their actions and beliefs are depicted as unproblematic. Contingent accounting, however, involves the subjective nature of scientific activity, incorporating social influences, personal idiosyncrasies, historical factors, craft skills and luck. Any potential contradictions and inconsistencies generated by the two repertoires are reconciled by the 'truth will out device' (Gilbert and Mulkay 1982).

Through the Myotonic Dystrophy research process, data were published in scientific journals. Scientific papers follow a conventional pattern; scientists present their data in papers as unproblematic, omitting issues such as mistakes. This serves to preserve the anonymity of the author and thus present the data in such a way that 'the research becomes "anyone's" research' (Gilbert 1976: 285). In this formal recounting of the research the empiricist repertoire was used. The knowledge claim was legitimated by making explicit that established methods, materials and procedures were used and any subjective elements, such as the characteristics of the author and the social nature of science, were eliminated. Thus the cloning of the gene was presented as an objective event based on experimentation and an ordered search for the gene. Through the mechanism of publication the Myotonic Dystrophy research network contributed to the body of knowledge about the Myotonic Dystrophy gene, first identifying the chromosome on which it lay, and later by establishing close markers which served to narrow down the area of interest.

Using the empiricist repertoire in both scientific papers and interviews the Cardiff scientists predicted that the gene would be located by means of good science, the unproblematic communication of data and the exchange of resources between the groups. Using this repertoire they also described the progress of the research and the techniques used by the Cardiff group to home in on the specific area and ultimately identify the genetic mutation. By 1991 the collaborative research had narrowed the gene location to a 200 kilobase area on chromosome 19. In order for the Cardiff scientists to clone this region a hybrid was constructed by a colleague at Massachusetts Institute of Technology. This hybrid was a mixture of rodent and human chromosome containing 10 kbs of DNA from chromosomes 17 and 19 which was treated with X-rays in order to fragment the chromosome into small pieces. Cells were then selected which had retained pieces of human chromosome close to the Myotonic Dystrophy locus. A restriction map of the region was created.[4] The library created contained two megabases of chromosome 19 which included the previously defined close markers. The human positive clones were sent to Cardiff and researchers there confirmed that they were human. In the process a probe, 59a, was pulled out; this included a repeat of DNA which it was believed could identify the Myotonic Dystrophy gene. The group had been working on this particular area of DNA for about a year: 59a was sequenced and found to be adjacent to the Myotonic Dystrophy gene. The polymorphism[5] identified by 59a was a 1 kb insertion. Another polymorphism showed almost total disequilibrium, that is one of the two alleles identified is associated with Myotonic Dystrophy to a greater extent than predicted from allele frequencies determined from unaffected chromosomes.

The London scientists also used the empiricist repertoire in accounting for their research. The rhetoric corresponds to the standard conventions for publications in the natural sciences (Bazerman 1988, Myers 1990). The paper they published in the journal *Nature* described the materials, methods and procedures used to map the area and identify the mutation (Aslandis *et al.* 1992):

> The Myotonic Dystrophy locus has been characterised both genetically and physically and recent efforts have concentrated on constructing a long range restriction map of the critical region. The distance between the genetic markers ERCC1 and D19551 is –600 kilobases (kb).
>
> Yeast artificial chromosomes (YAC) clones have been screened with the flanking markers ERCC1 and D19551 and a contig around the latter probe has been described. A more comprehensive cosmid and YAC contig has been constructed across the entire Myotonic Dystrophy critical region. Because linkage disequilibrium studies suggested that the probe D19S62 is closest to the Myotonic Dystrophy locus, YACs containing this probe were characterised.

Scientists also participate in informal discourse in interviews and at conferences when they may use both the empiricist and the contingent repertoires of accounting. When using the latter means of accounting, the scientists spoke of factors such as laboratory expertise and the differential use of techniques, of social factors such as competition, individual career trajectories and the social organisation of laboratories which they believed had influenced the progress and success of the research. By means of this repertoire the scientists emphasise a view of research stressing uncertainty. Through the two interpretative repertoires the participants in the Myotonic Dystrophy research could therefore weave the rhetoric of luck with skill, uncertainty with predictability and serendipity with incremental scientific progress. The scientists' accounts refer repeatedly to the element of competition, as well as collaboration and cooperation, between the various research groups that were involved in the search for the gene. The rhetoric of predictability and chance is often embedded in accounts of the differences between research groups, the research strategies they developed, and individual scientists' contributions to the overall breakthrough. The predictability of the eventual discovery of the gene is thus balanced with the contingent factors that determined precisely what was discovered, how, where and when.

Throughout the research, candidate genes were often identified. Once these had been identified it was necessary to establish whether they were the Myotonic Dystrophy gene, and progressively to eliminate sites from the search. Despite the prediction that the gene would be found, there had been considerable uncertainty over when this would occur. Previous experience with the search for the gene responsible for Huntington's Disease had shown that a precise trajectory for such a discovery was not predictable. The perceived progress of the research into Huntington's Disease was used as a moral tale. The research had been going on for many years before the gene was finally cloned in 1993 – after we conducted our interviews. The search for Huntington's had involved larger and better resourced groups, as well as a good intellectual input and technical skills. By 1991 it had become apparent to scientists working on Myotonic Dystrophy in both Cardiff and London that, although the search had been progressively narrowed to a tiny region, there remained many genes in the area – all of which could potentially have been the target. The uncertainty which existed over timing was articulated by means of the contingent repertoire.

One of the Cardiff scientists succinctly expressed a degree of uncertainty:

Int: Did you realise that they were homing in on the gene, that it was imminent in the weeks leading up to the discovery.

Sc: Not really, no. Everybody knew it was all very close. The precise way in which it came about was a bit unexpected.

In the following, later, extract from the same interview the scientist talked about the nature of collaboration between the groups and the flow of information between them:

Int: In the weeks or months leading up to the cloning of the gene, do you think there was more withholding of information as people felt they were getting there?

Sc: No, because there had been a very prolonged period of getting there, people had known they were more or less at the gene for a very considerable period and it was essentially only when the unstable fragment was identified that people had confirmation that they had reached where they already thought they had, and that all happened very suddenly and very rapidly . . .

The scientists emphasised the amount of work involved in the research. As one of the London group put it: ' . . . we felt we'd been lucky enough to have the lucky break and that reflected all the work we'd put in. A lot of other groups had put a lot of hard work in too and we felt very lucky to have that lucky break, but then again we thought we didn't deserve it any less than anyone else'. They also expressed the belief that the gene could have been cloned earlier but that this would have been entirely due to chance. As one of the past members of the Cardiff research group put it, ' . . . it's only a matter of time until you find it. It's an element of luck too – both I think. It's work, you have to do a lot of work and be lucky'. One Cardiff scientist talked to us – 'with hindsight' – about the particular contribution he had made and the possibilities it contained. This aspect of his work was concerned with a particular gene probe – labelled 59a.

Int: What was the time scale of this? I know you were working on it throughout, but when did you actually pull out 59a?

Sc: This is probably the embarrassing bit. Obviously I originally pulled out these twenty clones early on after about a year probably, eighteen months, something like that over the next . . . year.

Int: Would that have made it 1991?

Sc: Probably early 1991, and over that year I was working each of them up alongside finishing up the other typing and by the time I went to the European Society meeting in Leuven in the summer of ninety-one, we knew that 59a was quite strongly associated with Myotonic, but it wasn't, you know, enough to base a very intensive study on that particular region, although we were doing that sort of in the background, and obviously our collaborators knew about 59a, but the main data presented then was the D10 data [a different probe].

Later in that same interview we returned to the topic of probe 59a:

Int: Did you feel that if it had been left up to you, you would have preferred to concentrate on 59a, or did you feel it was the right thing to do to go in the directions that were suggested?

Sc: Interesting question. In hindsight, it would have been nice to follow the whole 59a through. At the time it didn't bother me that I was channelled into the other jobs as well because I was still working on the 59a stuff in a way, and getting mileage out of that . . . and also it was a group effort so you can't just have everything to yourself . . .

. . .

Sc: Well, I am very much afraid of blowing my own trumpet but I suppose it was a very significant step, it did very much narrow our concentration on a particular region. It did narrow our concentration and the data didn't seem to conflict with anything previous, which made us sure about it.

Int: . . . Do you feel it was a significant discovery in its own right, 59a?

Sc: Yes I do. This is the accepted way of finding these things, and that is the protocol I followed and the fact that I found the cDNA that was as near as damn it next to the gene, is very good. Makes me feel quite good about it.

Here, then, the scientist mingles a number of thematic elements in the account. It starts with 'embarrassment' – because the potentially fruitful line of research was not pursued vigorously – and ends with 'blowing his own trumpet', identifying the personal role played in the overall work of discovery. The mingling of self-deprecation and self-praise is thoroughly characteristic of scientists' discourse (Mulkay 1991: 169–82). The contingent timing of the outcome, and the decision not to concentrate on work with the probe, is contrasted too with 'the accepted way of finding these things': the empiricist and the contingent are closely interwoven. The same interview continues:

Int: Do you feel the discovery could have been made earlier?

Sc: Well, you always feel that, don't you? We had 59a for a year or so, doing various things with it, but unfortunately we had never got round to doing that particular thing before.

In much the same vein, another of the geneticists from Cardiff – subsequently at a different UK University – stressed the temporal uncertainty:

Int: Did you realise as time went on how close you were and that the discovery was going to be made?

Sc: Well, we always knew it was very close. The job that I have now – I actually was appointed a year before I took it up, but I asked to stall as long as I possibly could, because I knew we were very close. The problem was that we had been close for a long time, and sometimes with hindsight it was obvious that's how it work. But for example, Huntington's Disease they are very close – in a similar position to what we were two years ago I would say – and they could find that gene anywhere from tomorrow to two years on, or even longer depending on the nature of the mutation.

Int: Was that the same with the Myotonic gene – it could have been found any time?

Sc: It was entirely chance, although there was a lot of hard work and we can now look back and say OK it's obvious that we were about to jump over the edge and we were going to find it, but at the same time when you were busy working away there it didn't seem that obvious. So the reason everything evolved as quickly as it did was because of the nature of the mutation. Now had it been a point mutation, like say just a single base pair change, we would probably still be looking.

Although the scientists knew from their data that they were very close to locating the gene there was uncertainty regarding just how 'close' they were: although the breakthrough was imminent it was impossible to know when it would occur. The long process of research was, therefore, juxtaposed with the uncertainties and short-term unpredictability of the actual discovery. For instance, one of the senior Cardiff scientists accounted for things in the following way:

Int: So did you realise throughout the research that you would discover the gene?

Sc: No, it just goes on for so long, and because the Huntington's gene had taken so long, and they had bigger groups and much more high powered groups. . . . There's all these big groups and they hadn't cloned the Huntington's Disease gene after so many years, so I thought it was going to take us at least as long as it had taken them.

Int: But did you feel you would get there?

Sc: I wasn't sure. Initially I thought we would and then as time went by and the others were catching up because they'd got this large scale cloning, contig-building experiment going on, that it was just going to be luck then, because they stood as good a chance as us and once they'd managed to localise our clones on their map they wouldn't need our information, they could generate their own information. . . . Whereas we'd taken a semi-organised approach, based on more biological observations, rather than just doing blanket experiments we were at least trying to adopt an intelligent approach and go for little bits, rather than go for the whole thing.

Int: So there was an element of luck or chance at the end or was it that you were being more systematic?

Sc: Well, yes but I think there is with anybody, because if you're looking for genes, you can narrow it down to a very small region, but there's still going to be a lot of genes in that region, so it is luck.

Another of the senior members of the Cardiff group also expressed a similar contrast between the imminence of the discovery and the uncertainty of its actual appearance:

Int: Did you realise as the research was going on how close you were?

Sc: No, we knew we were fairly close because of the disequilibrium data, but we weren't quite sure exactly what it meant because we'd got two regions we'd been looking at which both showed strong disequilibrium, one of which was much stronger than the other region, but because of the mathematics involved we weren't quite sure whether what we were seeing was a real effect and I always tend to err on the side of caution, so while we thought we had something, the gene we'd been looking at which was the gene, the only gene we knew about at the time, 59a, we hadn't found any obvious changes in that gene to suggest that it was the Myotonic Dystrophy gene, we weren't quite sure what was going on.

Another Cardiff scientist constructed a very similar account of imminent discovery and unpredictability, linked with the success of good scientific work:

Sc: . . . The [overseas] group rather ran into the sand. They were clearly not going about things in the right way, with the result that they were the only group in the original collaboration that weren't involved at the end – which was sad but, actually, probably appropriate. Whereas I think everybody would have felt if any of the other groups hadn't had a stake in the final discovery it would have been unjust. Because all the other groups had in one way or another got very close to things, and had actually shared a whole lot more.

Int: You mentioned the Washington meeting in the October of 1991 – was that when people started to believe that it was imminent?

Sc: No, people had been believing it was imminent for probably the best part of a year before, and in fact since the neuromuscular disease meeting in Munich, which was just about a year before, and at the Washington meeting people were rather frustrated because of having thought it was imminent. Nothing had happened – well, quite a lot had happened – but there weren't any results basically, so people knew that something must happen, but it was very unclear what was going to happen.

Int: So could the gene have been found earlier?

Sc: Yes. Well. Could have, but it would have been a rather a chance process. It was a question of what people might have happened on before, yes. It could have been cloned quite a bit earlier from our own lab's involvement which had actually cloned part of the gene [59a]. But because it was not part of the gene with the mutation in, it wasn't possible until later to prove that indeed it was part of the gene.

Here and elsewhere, the interviewer and the scientists use the terminology of 'closeness'. In the context of this particular kind of research, closeness is a particularly apt usage. The term, of course, refers to the imminence of the discovery – implying a progressive approach to the right answer. For the scientific work involved in finding and cloning a gene, closeness also helps to capture the progressive mapping of chromosomes and approximation to the precise location of the gene. The spatial metaphor for the discovery process is thus especially appropriate and evocative.

In using the empiricist repertoire the scientists emphasised the progressive nature of the search. When using the contingent repertoire they also spoke of the element of luck which many believed was involved in the research. This chance factor was not expressed in terms of whether or not the gene would be located, for all the scientists interviewed believed that this would occur eventually, but in terms of when this would occur and in terms of which group would be the first to identify the mutation. As one of the senior members of the London group expressed it:

Int: I was wondering with people using different techniques how straightforward it was for them to replicate the results.
Sc: Well they had all the clones we had, we all had the same chromosome and it was just which bit you picked up and looked at. We had the lucky break – we looked at the right bit first . . .

Another of the scientists employed the vocabulary of luck in a very similar way – emphasising one unpredictable element of genetic work:

> There is always an element of luck in a sense that you may have a series of genes that you are looking at for fragments of DNA and if you start at the wrong end and it's the one at the other end, I mean it's as simple as that in that sense, but if somebody starts at the other end they get it first and you don't and it's just sometimes it can be a matter of chance as to which way you do things. So I think there's always an element of that and you won't get away from that, because a lot of it you have to be methodical, you don't want to miss things, so you design whatever method or process that's good for you. I think there will always be an element of luck for certain . . . even if you start with the same material there's just no way of knowing when it's going to happen exactly.

Methodical work and chance are closely juxtaposed in this description of scientific discovery. They are closely associated more generally in these accounts. One of the more junior members of the London group expressed things in terms of surprise – again, set against a background of temporal uncertainty.

Int: Did you realise how close you were in the weeks leading up?
Sc: In August, I had just decided, by then it was becoming apparent that there were quite a lot of genes there and any one of them could have been the Myotonic Dystrophy gene, so it looked as if it was going to be another two years for all those to be searched for mutations, and it looked as though, it was like, Oh well, best finish up, got enough for a PhD to just leave it there really. But I had just done this last experiment to see . . . how it was being inherited, just in one

Myotonic family – just to have a look, so I could say 'Oh it looks as though this is the candidate gene among the other candidate genes at the time'. And when I pulled out the autorad I thought I had done something wrong, because there were all these funny extra bands, and that was a really small family, there were just two parents and three children. And I thought, this is really strange – it looks like either there are some funny variants coming in, or I have just done something rather wrong. . . . So I picked an extended family of about twenty members, and about eight of them were affected, and when we developed that it was obvious to see that there was something. This funny extra band was definitely segregating the disease, and at that point we did get really excited.

. . .

Int: So what was your reaction then?

Sc: Well, we were really excited, but we couldn't quite believe it. We couldn't quite believe that nobody else had come across it, this gene, so far. The first thing we had done, we had probes available for all the other genes that we had pulled out, so we checked it against all of those, and it appeared to be a completely new gene that nobody had found so far. So that was really exciting.

In common with similar discovery narratives, individual scientists described their own individual emotions of surprise and excitement. Such accounts emphasise the individual's personal agency in the context of group work and collective progress towards solving the research problem. For instance, one of the London group described a critical moment in the discovery process in these terms:

Int: Did you anticipate locating the gene? Did you know that you were going to find the location of the gene in your research?

Sc: We were hoping to! I'm not sure really, because the groups were working on a collaborative basis. There is some competition as well obviously – I think mainly because everybody is grant funded and it goes on your track record. So the more publications you have, the better chance you have of getting a grant, and a lot of the recent techniques seem to come through the Americans, such as Cystic Fibrosis and Neurofibramatosis, they found the gene there. So I thought maybe we'd never get it ourselves, because the Americans seem to have such a large resource there, and money and everything like that. But we were hoping we would and I suppose we thought we stood a good chance because we had all the resources we needed to try and get there. It was just who was going to get there first really.

. . .

Int: Did you know how close the other teams were?

Sc: I'm trying to think when the last meeting had been. We knew everybody was in the right area, and everybody had clones from the area, and everyone was busy characterising these new clones and everything. And I think it just happened that we almost knew what we needed to look for – the differences between the patients and normal people – and I don't know whether it was completely luck or whether it was fortuitous, but because we worked towards this, it just happened that we found it like that. All the other groups had clones in the area which they could have done the same thing with, so I suppose we were first to get there.

. . .

Int: You said Y was the one who actually pulled out the autorad, what was the feeling then amongst your team?

Sc: We just couldn't believe it, because first of all we'd had a lot of problems . . . we were using probes in the area, we were picking up cDNAs which weren't on the clone.

Int: Picking out?

Sc: cDNAs, the genes, for other markers. And we found a lot of them weren't on chromosome 19, so when Y did the experiment and picked out the gene and mapped it, apparently it was on chromosome 19. So we were really excited at the start. And then we found it was polymorphic, it was informative, so we thought we'd got a new marker, so we were really excited about that!

Int: You were excited anyway!

Sc: Yes, that's right. We actually had a cDNA for chromosome 19. We were really chuffed, and I just kept bringing out new autorads and saying 'Look at this, look at this'. And then we ran . . . a comparison against patients and normal people. So we had been on the lookout, we had these filters ready made up to look for any differences within the gene we'd found which we'd isolated from the area, so we just couldn't believe that it had been so quick in finding it – cDNA – and on chromosome 19, it was polymorphic and it showed a difference. It was just incredible, because we kept thinking there's got to be something wrong, something's going to happen, we are going to find it's not the right one or something like that. We kept thinking that something was going to go wrong. We repeated it, confirmed it, and we just couldn't believe it.

As can also be seen from this particular account, surprise was expressed that British groups had found the mutation before the overseas groups. It was echoed in several of the other discovery narratives. Despite the collaborative nature of the research which had ensured that all the groups had reached a similar stage in the research, and that any time lag between the groups in the progress of their research was short-lived, the scientists believed that the resources available to the North American groups in particular had given them an edge which could have allowed them to clone the gene ahead of the UK groups. The scientists also referred to chance which they felt had allowed the overseas groups to replicate the results. The Lawrence Livermore group had also used the restriction enzyme Eco R1 to characterise their clones. Had they identified the polymorphism with a different enzyme it would have been necessary for them to digest all the clones with Eco R1. This chance use of the same enzyme allowed the group to perform a computer search in order to identify the particular clone.

In their contingent accounting for the discovery some informants also spoke of the bench skills which scientists developed in the course of their scientific careers. Several of the Cardiff scientists felt that skill at the bench was a factor which had affected the pace of research into Myotonic Dystrophy. They felt that researchers' technical skills varied enormously, that to achieve successful experiments one needed to be methodical and accurate and that inefficient science had delayed the Cardiff research. In predicting which group would locate the gene first they contrasted the scientists' skills and techniques with luck and chance. In the empiricist accounting of the nature of the collaborative relationship they described the regular communication with the other groups in the collaboration and the exchange of resources as unproblematic. They also described the various approaches that the groups adopted in their research. The American and Nijmegen groups were using what was perceived as a methodical approach to the problem. In their contingent accounting the scientists saw Cardiff's 'semi-organised approach' (as one of them described it) as more 'intelligent' than the more mechanistic approach of the overseas collaborators. On the basis of these perceptions the Cardiff scientists claimed to have predicted that some groups were less likely to locate the genes than others and this perception influenced the dissemination of resources. Probes, for example, were supplied to one group who were described as searching haphazardly: '. . . they weren't really considered to be a threat, and I mean it seemed pretty likely, I always felt that it was either going to be us who got it or X'.

Members of the London group emphasised the element of chance which had allowed them to clone the gene before the groups outside the United Kingdom. The small London research group was unable to undertake the same research strategies as the larger overseas laboratories. They were therefore more selective and felt that their tactic represented a gamble which paid off. As two of the London members told us, in the course of longer accounts:

> Sc: . . . I think that anyone working on the disease would have been able to get it . . . Although it was a fairly technically difficult strategy we took, it was a short cut and there were no guarantees and it came up trumps . . . I think any group could have made it given an unlimited time.

Sc: . . . We didn't have as many resources as other groups to actually look for the gene, so we had to use what we had available and in a way that paid off for us. Because we didn't have a megabase of genome to screen, we only had these YACs and a few cosmids.[6] It could have easily worked the other way, we could have got the wrong YACs or the actual experiment might not have worked.

Throughout the research on Myotonic Dystrophy the scientists expressed certainty that a gene which caused Myotonic Dystrophy existed due to the inherited nature of the disease. However, uncertainty was expressed by scientists over the nature of the genetic mutation. There are various mutations which can occur within genetic disorders; these include point mutations, insertions where additional nucleotides are present and deletions where part of the gene is absent. Genetic mutations can alter the processes of transcription and translation and so affect the protein synthesis, causing the disease symptoms or phenotype. The Fragile X gene had recently been located by scientists in the US and Holland. This disease causes learning difficulties and psychological problems in boys and, like Myotonic Dystrophy, is characterised by anticipation. The genetic defect of Fragile X was found to be an unstable repeat sequence of DNA which comprises 3 nucleotides, cytosine, guanine and guanine (CGG), and because of the shared phenomenon of anticipation, this influenced some of the scientists' beliefs about what form the mutation could take:

Sc: . . . I had been plugging the idea that with anticipation Myotonic Dystrophy was very similar to Fragile X and therefore it was well worth looking for some similar mechanism and therefore because an unstable repeat sequence had come out just six months before in Fragile X I had already suggested to be on the look out for something like this.

One of the London research group also gave us a vivid account of the epiphany:

Int: Did you realise how close you were to making . . . ?

Sc: Yes, I think by then we did. We were quite excited about these cDNAs, or what we thought were cDNAs we had pulled out, and knew that these experiments were ongoing. And as soon as I got back and saw the results, then I realised. 'This is it, finally a result that we had hoped'. I think the penny dropped for me in about May of ninety-one. I went to a meeting at Cold Spring Harbor and they were talking about this Fragile X mutation, and suddenly the penny dropped that this was what the Myotonic mutation would look like as well. I mean OK I could be wrong, but I expected to see something like that.

Conclusion

There are numerous sites for the sociological analysis of medical knowledge. The late modern division of labour in medical settings is highly complex and varied. Clinical expertise is no longer encompassed in the 'clinical gaze' at the patient's bedside, and clinical practice is no longer a craft based on signs and symptoms. The laboratory is increasingly central both to the production of new medical knowledge, and to the production of more routine diagnostic information. The sociology of medical knowledge cannot be confined to the work of clinicians, nor simply to the clinical encounter at the bedside or in the consulting room. The rapidly changing field of medical genetics has provided a case in point.

A great deal of knowledge production germane to clinical practice takes place in the laboratories and other 'backrooms' of medical institutions. Sociological analysis cannot confine itself only to the social construction of clinical entities and diagnostic categories in the course of clinical encounters. One cannot afford to take for granted the discovery claims and more routine knowledge that emerge from scientists' laboratory work. The priorities, opportunities and limits of clinical practice are set by the discoveries and technologies of laboratory science. As Fujimura (1987) pointed out, scientific research is conducted within the bounds of normal science, and on the basis of 'doable' scientific problems or puzzles. The construction, therefore, or doable research – and hence the possibilities for clinically-relevant research – is strongly influenced by scientists' constructions of predictable research outcomes. Research groups and laboratories are established and 'tooled up' (with personnel, equipment, funding and local expertise) to pursue research along well-established lines.

The discovery of the Myotonic Dystrophy gene was regarded by all the scientists as a predictable outcome of the research being pursued at the different laboratories. The business of gene discovery was a well-established one: the techniques were well understood. The Myotonic 'breakthrough' was certainly greeted as a major discovery, and was regarded as a very considerable success by those who claimed the discovery, and by their peers elsewhere, but it is described by the protagonists in terms of a classic normal science model. The scientists had faith that one of the laboratories which periodically collaborated and competed in the race for the gene would in fact make the discovery. One possible construction of the discovery accounts, therefore, would treat the Myotonic Dystrophy gene as an entirely foregone conclusion. Equally, however, we have shown that the predicted outcome was not narrated in those terms. The scientists' accounts show the co-existence of different narrative registers. That is, their stories were constructed in terms of contrasting styles of explanation. They stressed the predictability of the discovery, and attributed success in making the breakthrough to hard work and skill. Equally, they stressed the elements of chance in determining the precise timing and place of the discovery. The themes of predictability and unpredictability were interwoven: while the discovery was regarded as inevitable, the precise location and character of the genetic anomaly were described as much less predictable. Despite the construction of predictability, the scientists' versions repeatedly stressed the conjuncture of contingent factors, and chance, together with competence and skill. Again, therefore, we have seen in these scientists' accounts a genre of mixed accounts. The interpolation of the second register of rhetoric accomplishes a significant function for the scientists' own constructions of reality. If success were to be accounted for totally in terms of normal science, and totally predictable outcomes, then there would be little place in the accounts for competition between research groups, with consequent notions of 'success' in the race for priority in publishing results. The register of skill and luck allows for the attribution of local success against the backdrop of normal scientific work.

While the genetic scientists could regard the general principles of the work as predictable, the research being conducted against a background of well-established laboratory techniques and a high level of consensus among collaborating and competing groups, this did not rob the discovery itself of significance. The genetic basis of a given disease may be treated as conventional wisdom, and a general location for the gene may be suspected, but it remains a major task to identify the location, and to describe the specific anomaly responsible for the disease. In the absence of such precise discovery work, there is no basis for clinical intervention. The long-term significance of the discovery discussed here lies in its consequences for clinical medicine as well as in the more basic scientific research that went into it. The identification of the gene responsible for Myotonic Dystrophy, and of the basis for increasing severity from generation-to-generation, was therefore greeted as a major breakthrough for medical genetics.

The geneticists' chronicles of their discovery suggest a more general feature of scientific work. It is, as we pointed out at the beginning of this chapter, conventional for 'internal' descriptions of scientific work and the logic of scientific discovery to emphasise the role of prediction. Prediction, hypothesis formation, hypothesis-testing and falsification are, together with replication, linked in most standard accounts of scientific knowledge. The genetic scientists' accounts suggest that prediction is far from straightforward. Just as the scientific 'discovery' itself is far from being an unequivocal unitary event, so the nature of prediction is complex. Commonsensically, one might think that prediction in science was necessarily a prospective issue – constructed about future events and subject to confirmation in the present. Our discussion here suggests that predicted discovery may also be a matter of *retrospective* accounting. Of course, given that our data here are retrospective narratives, we unavoidably emphasise that aspects of scientists' accounts. Nevertheless, the narratives we have reported on suggest that what 'counts' as the relevant prediction or set of predictions may be found retrospectively. The ambiguity to be found within the genetic scientists' accounts illustrates the extent to which there is, as it were, narrative leeway in the retrospective construction of prediction and discovery. The nature and temporal trajectory of the discovery are subject to narrative working and re-working. By the same token, the nature of the discovery itself is subject to *post hoc* accounting. As we have emphasised, the discovery is not an event in itself. The discovery is not merely a claim entered by particular groups or individuals, it is a process of accounts and constructions. The discovery claim itself must be entered and supported through various modes of accounting and representation, and its status *as* a discovery is subject to processes of definition and redefinition – on the part of the scientists themselves, and by others in the scientific community.

Processes and outcomes of medical discovery, then, demand attention from sociologists of scientific knowledge and from sociologists of medicine in equal measure. We need to understand the social processes and cultural forms through which innovations and 'breakthroughs' are accounted for. This present contribution deals with but one aspect of this complex area of representation. There are many sites of knowledge production and reproduction that deserve equal attention. The Myotonic Dystrophy gene story itself is and will be subject to multiple tellings: not just by the scientists concerned, and their collaborator/competitor groups, but in the academic and professional journals, the mass media, in medical textbooks and works of popular science. In none of these sites of knowledge production does a single true story reside. There are multiple versions, and the stories undergo repeated transformations. There is a collective need to address these narratives and representations if we are to grasp the multiple realities of medical knowledge. Myotonic Dystrophy is not a single entity in itself. It is assembled as a clinical entity or as a diagnostic category on different occasions. As we have seen in this account, it was reconstructed as a particular kind of genetic anomaly as a consequence of the scientists' discovery claims. That particular enactment of Myotonic Dystrophy in turn will have direct consequences in the reformulation of the entity in other sites of medical work and knowledge production. Likewise, the reformulation of the disease – from a genetic disease of 'unknown' causation – will have major implications for other parties and other interests. Medical charities, support groups and individuals will start to redefine the future: in terms of future research and funding, future treatments, and future generations. The clinical category of Myotonic Dystrophy, then, is distributed across a wide network of actors and agencies. Significant change in the definition of the disease at one point in that network will have far-reaching effects on the definitions that are sustained elsewhere.

Acknowledgements

We gratefully acknowledge the help and cooperation of the members of the research groups who participated in the study. The research was supported by the Economic and Social Research Council (Grant No. R000234102); the interpretations offered here are the authors' and do not represent the Research Council's policy. We are also grateful to the anonymous referees whose constructive comments helped us to prepare this chapter for publication, and to Sara Delamont for her reading of its successive drafts.

Notes

1 A probe is a cloned DNA fragment that recognises an identical DNA sequence.
2 An autorad, or autoradiograph, is a 'photographic' plate of DNA fragments.
3 cDNA is 'copy DNA', produced by cloning techniques.
4 A restriction map shows the positions of different restriction sites in a DNA molecule. A restriction site is a particular sequence of DNA that is recognised by a restriction enzyme, which cuts the DNA at that point.
5 Polymorphism: a gene that is polymorphic comes in many different forms (alleles) in a population, though no individual can have more than two of them.
6 YAC: yeast artificial chromosome – large fragments of human DNA can be made to resemble a yeast chromosome and can thus be cloned in yeast cells. Cosmid: a cloning vector used to clone large expanses of DNA.

References

Aslandis, C., Jansen, G., Amemiya, C., Shutler, G., Mahadevan, M. *et al.* (1992) Cloning of the essential myotonic dystrophy region and mapping of the putative defect, *Nature*, 355, 548–50.

Atkinson, P. (1995) *Medical Talk and Medical Work: the Liturgy of the Clinic*. London: Sage.

Barnes, B. (1982) *T.S. Kuhn and Social Science*. London: Macmillan.

Batchelor, C., Parsons, E. and Atkinson, P. (1996) The career of a medical discovery, *Qualitative Health Research*, 6, 224–55.

Bazerman, C. (1988) *Shaping Written Knowledge: the Genre and the Activity of the Experimental Article in Science*. Madison: University of Wisconsin Press.

Brannigan, A. (1981) *The Social Basis of Scientific Discoveries*. Cambridge: Cambridge University Press.

Brook, D., McCurrach, M., Harley, H., Buckler, A., Church, D. *et al.* (1992) Molecular basis of myotonic dystrophy: expansion of a trinucleotide (CTG) repeat at the 3' end of a transcript encoding a protein kinase family member, *Cell*, 68, 799–808.

Buxton, J., Shelbourne, P., Davies, J., Jones, C., Van Tongeren, T. *et al.* (1992) Detection of an unstable fragment of DNA specific to individuals with myotonic dystrophy, *Nature*, 355, 547–8.

Casper, M.J. and Berg, M. (1995) Introduction: Constructivist perspectives on medical work: medical practices and science and technology studies, *Science, Technology and Human Values*, 20, 395–407.

Clarke, A.E. and Fujimura, J.H. (eds) (1992) *The Right Tools for the Job: at Work in Twentieth-Century Life Sciences*. Princeton, NJ: Princeton University Press.

Collins, H.M. (1992) *Changing Order: Replication and Induction in Scientific Practice*, 2nd Edition. Chicago: University of Chicago Press.

Connor, S. (1993) The brave new biology, *The Independent on Sunday*, 31 January, 40–1.

Fu, Y.-H., Pizzuti, A., Fenwick, R., King, J., Rajnarayan, S. *et al.* (1992) An unstable triplet repeat in a gene related to myotonic muscular dystrophy, *Science*, 255, 1256–8.

Fujimura, J.H. (1987) Constructing do-able problems in cancer research: articulating alignment, *Social Studies of Science*, 17, 257–93.

Fujimura, J.H. (1988) The molecular biological bandwagon in cancer research: where social worlds meet, *Social Problems*, 35, 261–83.

Fujimura, J.H. (1992) Crafting science: standardized packages, boundary objects, and translation. In Pickering, A. (ed) *Science as Practice and Culture*. Chicago: University of Chicago Press.

Gilbert, N.G. (1976) The transformation of research findings into scientific knowledge, *Social Studies of Science*, 6, 281–306.

Gilbert, N.G. and Mulkay, M. (1982) *Opening Pandora's Box: a Sociological Analysis of Scientists' Discourse*. Cambridge: Cambridge University Press.

Harley, H., Brook, D., Rundle, S., Crow, S., Reardon, W. *et al.* (1992) Expansion of an unstable DNA region and phenotypic variation in myotonic dystrophy, *Nature*, 355, 545–6.

Harper, P. (1989) *Myotonic Dystrophy*. London: W.B. Saunders.

Kuhn, T.S. (1962) The historical structure of scientific discovery, *Science*, 136, 760–4.

Kuhn, T.S. (1970) *The Structure of Scientific Revolutions*, 2nd Edition. Chicago: University of Chicago Press.

Lippman, A. (1991) Prenatal genetic testing and screening: constructing needs and reinforcing inequities, *American Journal of Law and Medicine*, 17, 15–50.

Manadevan, M., Tsilfidis, C., Sabourin, L., Shutler, G., Amemya, C. *et al.* (1992) Myotonic dystrophy mutation: an unstable CTG repeat in the 3' untranslated region of the gene, *Science*, 255, 1253–5.

Marteau, T. and Richards, M. (eds) (1996) *The Troubled Helix: Social and Psychological Implications of the New Human Genetics*. Cambridge: Cambridge University Press.

Mulkay, M. (1991) *Sociology of Science: a Sociological Pilgrimage*. Milton Keynes: Open University Press.

Myers, G. (1990) *Writing Biology: Texts in the Construction of Scientific Knowledge*. Madison: University of Wisconsin Press.

Parsons, E. and Atkinson, P. (1992) Lay constructions of genetic risk, *Sociology of Health and Illness*, 14, 437–55.

Parsons, E. and Atkinson, P. (1993) Genetic risk and reproduction, *Sociological Review*, 41, 679–706.

Seidel, J.V., Kjolseth, R. and Seymour, E. (1988) *The Ethnograph: a User's Guide*. Amherst, MA: Qualis Research Associates.

5. Vital comparisons: the social construction of mortality measurement

Mel Bartley, George Davey Smith and David Blane

Introduction

The ideas behind the present chapter arose from work on the implications of different forms of measurement for health and prevention policies. There are a number of different ways in which the risk of illness or mortality is measured in epidemiology and applied in public health planning and health promotion strategies. Different methods may give different pictures of risk and thereby give rise to conflicting policy implications.

One recent example of some political importance may be taken from the debate on health inequalities. If the traditional Standardised Mortality Ratio or SMR[1] is used to depict the major causes of class differences in mortality risk, coronary heart disease appears to be the dominant cause. But another alternative, Years of Potential Life Lost (YPLL), shows a preponderance of accidental and violent death (Blane *et al.* 1990). The policy implications are sharply at variance. Yet there is no way in which a purely technical decision can be made between these forms of measurement.

There may be some help at hand, however, from the social study of science and technology (SST). In the various versions of this perspective, facts and artifacts are regarded as 'socially shaped', and methods of measurement are expected to reflect the interests and objectives of the social groups involved in their development as well as the state of the 'real world out there'. Some sociologists and social historians of science and technology have begun work to arrive at an understanding of these processes. Such work shows that it may be enlightening to open the black box of epidemiological measurement and seek to understand the ways in which such methods of measurement arise.

We offer here a preliminary attempt to use the 'translation approach' (Latour 1987, Coutouzis and Latour 1986) for the social study of science and technology as a methodological guide to understanding the origins of potentially conflicting accounts of health problems. It places the production of facts and artefacts within wide-ranging social networks, and provides a set of orientating concepts for studying these networks. The translation approach suggests that it is when strong alliances are stabilised between widely different parties – including scientists, funding bodies, branches of governments and industries – that 'hard facts' and 'good methods' emerge. An understanding of the strategies which produce accounts of the risk of disease and death may help to clarify some of the difficulties which arise when we attempt to use epidemiological evidence in public health policy making.

The aim of the chapter is to suggest a possible agenda for future studies rather than to settle any questions. This means that we have not dealt with any of the measures in great detail. It is only possible to write a chapter like this one because detailed case studies of some of them are already available (Eyler 1976, 1989, Higgs 1991, Szreter 1991). The objective here is to draw out some more general significance from these case studies, using a specific theoretical and methodological approach, as an invitation to others to discover new examples and analyse them more fully. As 'participant observers' of public health research, rather than historians or science-policy analysts, we cannot hope to do more than this. We undertake the task because we feel that the significance for the public health sciences and thus for health policy and planning of the existing studies is not sufficiently acknowledged.

From the outset it must be clear that we do not mean to exclude 'states of nature' from the analysis. The reason for doing social studies of science is not that nature does not exist, but on the contrary because of the infinite variety and complexity of the natural world. Choices have to be made as to which of an infinite number of possible questions to pursue, how out of an infinite number of possible ways to pursue it, and how many resources to use. Understanding these choices is the task for the social study of science and technology. The state of science and technology at any one time is not the outcome of social activity as the result of some kind of conspiracy; it is so because it could not be otherwise.

Theoretical and methodological perspective: the translation model of science

An advantage of the translation model for studying scientific debates is that it lays down not only general principles but also, crucially, a clear set of rules of method (Latour 1987, Coutouzis and Latour 1986). The most important ones for our purposes here are:

1. *Study scientists in action rather than looking back over the histories of established ideas.* In this way the method is essentially ethnographic. We look at the way in which scientists use statements about the 'correct methods of measurement' to enrol other groups into alliances through which they may pursue their objectives.
2. *Do not make up your mind beforehand who is a real scientist and who is, say, a politician or a bureaucrat or a businessperson.* Fact-builders come in many guises, and it is only after the event that facts look as if they had

been created purely by people in white coats. To understand the construction of fact it may be necessary to look far outside the circle of scientists alone.

3. *The principle of symmetry*: This warns against taking sides when studying the type of controversy which accompanies new ideas in science and technology. We need to put ourselves in the shoes of contemporaries who do not know what will turn out to be the truth about, for example, what the size of class differences in mortality is; whether cigarettes cause lung cancer; or whether Liverpool is more or less healthy than Norwich.
4. *The fate of facts lies in the hands of their users.* When we look at claim and counter claim in a scientific debate, we can see that there is no other way in which such a dispute could be decided. The notion of the decisive experiment or observation is discredited by the 'experimenter's regress' (Collins 1985). This is the paradox that if doing an experiment for the *first time ever* it is difficult to know whether a result which is very different from your expectations is due to nature's being different from what you expected or to error, or, say, faulty equipment. To some extent the quality of the equipment, the skills of the experimenter etc. are themselves judged according to the outcome of an experiment. But if we admit this we are into a circular argument: the experimenter's regress. Disputes end when those who proceed as if one claim were true gain precedence for a variety of social and political reasons. It is not the truth of claims which decides their fate but the fate of claims-makers which constructs truth.
5. *Science is not created in individual minds but depends crucially on networks of people.* An idea will only become a fact if these networks use it. At the centre of such networks, and important influences in creating and sustaining them, are often what Latour calls '*centres of calculation*' (Latour 1987, 1988). We should therefore be on the lookout for such centres as important if inconspicuous participants in scientific controversies.

There have now been several studies examining the construction of 'cause of death' (Prior 1985a, 1985b, Bartley 1985, Armstrong 1986, Thevenot 1987) and of measures of mortality *risk* by sociologists and social historians of science. The most notable of these for the present work is Szreter's (1991) account of the early relationships between the General Register Office and the public health profession. We intend to draw upon this case study of the GRO's use of the Standardised Mortality Ratio (SMR) to try to generalise its account of the ways in which new methods of measurement are accepted; using Doll and Hill's (1950) introduction of the odds ratio as a further example. We shall examine the associated controversies about how to measure mortality risk according to the methods of the translation programme: giving equal importance to everyone who was involved; looking for the establishment of centralised methods of gathering observations and homogenising them into general fact; examining the formation and dissolution of alliances between social groups; and describing the ways in which one method rather than another has come to be taken up and used as the correct one, i.e. taken to be giving the true picture.

Comparing vitality: from average age at death to the Standardised Mortality Ratio (SMR)

What is the average length of life?

When looking for the reasons why age structure eventually did become built into the methods of measurement used by public health experts, we began to examine the debates in which they were engaged. The rules of method of the translation programme demand that we be aware of wider networks, and also remind us of the possible importance of 'centres of calculation'.

The Standardised Mortality Ratio is well established as the method for comparing health advantage and disadvantage in vital statistics. However, looking back to the beginnings of the measurement of mortality risk shows that this method of measurement was only accepted after many decades of often acrimonious debate. By following the protagonists of the debate, and the development of institutional and organizational links between them, we may be able to gain a better understanding of why the SMR eventually became so well established.

A method for measuring life expectancy was needed by early social reformers concerned with the consequences of rapid social change in 19th-century Britain (Goldman 1991). The conventional method for estimating population mortality and life expectancy at that time was explained by Edwin Chadwick in the following way:

> The mode generally in use is to take the proportion of deaths to the population . . as representing the average age of death in any population (Chadwick 1844: 1).

According to this method, if one in 22 of the population died annually the average duration of life for all who died was 22 years. Chadwick ridiculed this reasoning. He favoured adding up the ages of all who died and taking the mean. Chadwick's 1844 paper to the Statistical Society of London was mainly concerned with the best method of collecting data which would allow the direct calculation of average ages of deaths. For example; this could not be done without reasonable certainty that all deaths were being registered, and age given accurately. Complete registration at the time was, in fact, a recent innovation: deaths of people who were not members of the Church of England were only registered after enactment of the Registration Act 1836, which came into force on 1 July 1837. This legal and administrative change made it possible to estimate life expectancy in a new way, although Chadwick complained that registrars were not sufficiently careful when noting the age of the deceased.

The 'mean age at death' measure had the virtue of not requiring knowledge of the size of the population in each age group, but only the numbers and ages of decedents. Calculation of death *rates* in each age group could not be credible until knowledge was available of the numbers in each age group in the whole population. In the first national Census of 1821 some attempt was made to record age, but it seems to have been rather a sensitive subject, and was only obtained by asking people to indicate where they fell in a list of five-year age groups – apparently a method rather similar to the way in which modern surveys ask about income (with a 'flash-card'). Enumerators were instructed to ask the question only if it could be done in a way which did not embarrass or offend respondents. There was no age question in the 1831 census (Nissel 1987).

Up to 1841, therefore, the only alternative to Chadwick's method of estimating life expectancy had been based on life tables. Because life tables require death rates in each age group, they can only be compiled where there is knowledge of the numbers and ages of the population. Such information was collected in local censuses, which had been periodically undertaken during the 18th century, *e.g.* in Carlisle, Glasgow, Chester, Northampton and on the continent in Montpelier and Geneva (Chadwick 1844, Greenwood 1946). These studies seem to have been undertaken for the life insurance companies. They did have access to local censuses giving numbers and ages of the population, and could therefore calculate the proportion in an age group that would live a further five years, 10 years, and so on.

Under the Population Act of 1840 the Registrar General was made responsible for the Population Census as well as civil registration. The national Census data on the numbers and ages of the population were centralised into the same General Register Office (GRO) which collected certificates of births, marriages and deaths: thus establishing a powerful 'centre of calculation'. So the newly available national statistics of mortality and age structure had extended and made permanent the local and episodic data collection exercises of the 18th century. And the people who knew a lot about the use of this kind of information were not public health specialists or reformers, but actuaries.

The conflict of views on the measurement of vitality between public health and actuarial science is illustrated in Chadwick's debate with F.G.P. Neison at the Royal Statistical Society (Chadwick 1844, Neison 1844). Neison is not a well known figure in the history of public health. He was an actuary, employed by the Medical Invalid and General Life Office. He agreed that insanitary conditions, such as those which Chadwick's Report described in Manchester, should be abolished but for ethical rather than medical reasons. Neison argued that Chadwick could not have demonstrated an effect of insanitary conditions on mortality because the method which he used for measuring life expectancy was flawed.

Chadwick (1844) had argued that insanitary conditions were shown to be unhealthy by the lower (as calculated by his 'average age at death' method) average life expectancy in such areas: people living in dirty towns died younger. Thus the average age at death in insanitary Bethnal Green was 25.8 years, while in Kensington it was 32.4 years. Similarly, the average age of death in newly industrialised Manchester was 22.9 years, compared with 38.4 years in the old market town of Hereford.

At the following meeting of the Statistical Society, Neison retorted:

> That the average age of those who die in one community cannot be taken as a test of the value of life when compared with that in another district is evident from the fact that no two districts or places are under the same distribution of population as to ages (Neison 1844: 41).

In other words, there might have been many deaths of young people in a community simply because there were many young people. He illustrated his argument by re-calculating Chadwick's average ages of death in a manner which took account of the differences in the age distribution of Bethnal Green and Kensington, Manchester and Hereford. This method greatly reduced or even eliminated the differences which Chadwick had claimed to show were associated with differences in sanitary conditions.

Neison was not surprised to find that life expectancy was no higher in Kensington than in Hackney. Following the work of Guy (1846), he accepted that the aristocracy and peerage, due to the 'dissipation' of their lifestyle, suffered a higher than average mortality. Drawing data from Friendly Societies, he was able to conclude that the healthiest social group was that segment of the industrious working classes who were members of such societies:

> . . . the general principle which seems to regulate the mortality of other classes, namely, that the humble but industrious working classes, whose prudential habits lead them to become members of these (Friendly) societies, are subject to a lesser rate of mortality than any other, and that the higher the class of society over which the observations extend, until the peerage, or highest class of all, is observed, in which there is less of the regular and healthful daily exercise essential to the condition of the industrious workman, the greater the rate of mortality; and for intermediate classes, a varying degree of mortality is observable, following pretty closely the scale of their position in social rank (Neison 1845–6: 292).

From a present-day vantage point we might dismiss Neison's reverse class gradient, on the grounds that it ignored the great majority of the working class who were not members of Friendly Societies. His advocacy of age and class standardisation has, however, now become a *sine qua non* of respectable reports on population health. At the time of the debate, Neison's technically damaging critique had little effect on the steadily growing acceptance of Chadwick's *Report on the Labouring Population*. It is not therefore the political influence of Neison's method which explains its persistence. Today Chadwick's aetiological hypotheses ('poverty and poor environment are health risks') are perhaps more accepted than they were in his own time – there have been no attempts to revise the picture of widespread premature mortality in unsanitary towns during 19th-century urbanisation. But the method on which this picture was based is now regarded, not so much as discredited but rather as a fossil in the prehistory of statistics: no comparison of mortality or morbidity between areas or social groups would today be undertaken without some form of correction for age structure (for example, for mortality see Drever and Whitehead 1995, and for a similar method applied to rates of morbidity see Charlton *et al.* 1994).

In reviewing this episode the importance of alliances, and, in this case at least, of the emergence of a new 'centre of calculation', is evident. However we may also notice that the networks which build facts need not be the same as those which build methods, and that the relationship between the two is complex and variable. A fuller understanding of how and why Chadwick's construction of the causes of ill-health in mid-19th-century urban centres came to be as firmly accepted as it is today might be an interesting topic for further study.

The next stage in the emergence of age standardisation as the appropriate method for describing health differences between areas and social groups is described in the next section.

The management of population: Farr's Biometer versus the Darwinists

The setting up of the GRO, and the centralisation within it of both civil registration and the Census, changed the nature of the networks within which public health debates took place. In the early 19th century, a whole range of such 'centres of calculation' were being established (Macleod 1988): not only the GRO but organisations such as the Alkali Inspectorate, the Passenger Acts and Factory Acts and their corps of inspectors.

Measuring mortality was something in which a range of different groups became involved, both doctors and others. What vital statisticians had to offer the fast-growing Victorian state were methods for bringing disparate types of information together and finding ways to turn them into aids to government. The most successful form of homogenisation in industrialising societies was to attach numbers to everything (the term 'statistics' is derived from 'state', and at this time statisticians were also referred to as 'statists' (Cullen 1975)). Statistics, in this analysis, is a perfect example of a device which can at the same time summarise a great mass of collected traces and, unlike mere head counting, preserve some of their variability. Latour sees this objective as shared by all 'centres of calculation':

> . . . an *equation* ties different things together and makes them equivalent. These activities go on in centres of calculation and are only significant as long as the networks that hold centre and peripheries together *do* hold (Latour 1988: 35, emphasis in original).

There is a view among some historians (*e.g.* Roberts 1959) that 'inspection' was the Victorian state's compromise between a desire for central control and the fear of over-centralisation. Seen in these terms, the observational role of the GRO becomes even more significant. And we do not have to look very hard at the writings of William Farr to find these views explicitly expressed:

> Statistics . . . the science of men living in political communities, was never in such demand as it is in the present day . . . politics is no longer the art of letting things alone, nor the game of audacious Revolution for the sake of change;[2] so politics . . . has to . . . call in the aid of science: for the art of government can only be practiced with success when it is grounded on a knowledge of the people governed, derived from exact observation (Farr 1872b: 417).

And in his Supplement to the 35th Annual Report to the Registrar General Farr observed:

> Progress will be accelerated by new methods in the sciences . . . by the application of calculation to all branches of human affairs (GRO 1875: xix–xxx; quoted in Humphreys 1975: 135).

The political was by no means the only consideration at stake, however. Inaccurate estimations of population characteristics also had serious economic consequences. We have seen that another important actor in the story of measurement of mortality risk was the insurance industry. Accurate life tables were essential for the calculation of premiums and the cost of annuities (Goldman 1991, Higgs 1991). As early as 1833, the actuary of the National Debt Office had complained that the defects of current civil registration of births and deaths made it impossible to run either government annuity schemes or the newer Friendly Societies. A miscalculation of life expectancy by Price, a founder member of the Equitable Insurance Society, using parish registers, had led to artificially low premiums. His view of life in towns was that the major cause of high mortality was

> the irregular modes of life, the luxuries, debaucheries, and pernicious customs, which prevail more in towns than in the country (GRO 1847: xiii–xiv),

and even more so among the better-off who were customers for annuities. Price's calculations were even adopted by the government, which raised money by selling annuities. Because life expectancy was underestimated, this resulted in the government losing large amounts of money (Humphreys

1975: 153–4). How could lives be insured properly, let alone profitably, if no one knew what life expectancy was (Farr 1837–8)? Farr pointed out that

> millions of money [had been invested] . . . on the principle that the duration of vital phenomena [*i.e.* the length of life] admits of calculation (Farr 1837–8: 703).

Parish registrars were neither reliable nor accurate witnesses: some were known to have falsified records and many registers were kept sloppily (Eyler 1976: 42–5). Against such contingencies, the traditional fear that if a Census became public, and included age, military competitors might be encouraged in their ambitions began to be questioned (Eyler, 1976: 39).

The public health doctors and actuaries found a third ally in religious dissenters. As long as registration was by parishes, non-Anglicans could not obtain certificates of birth, death or marriage enforceable at law. All these influences lent weight to the campaign for civil registration. As a result in 1836–7, as we have seen, a civil registration scheme was set up. Superintendent Registrars were appointed by all Poor Law Unions, usually the Clerk to the Board of Guardians; by 1838 superintendents had been appointed in 619 of the 626 Unions (Nissel 1987: 14). These Unions had in turn of course resulted from Chadwick's labours in the reform of the Poor Laws. Each Union had registration subdistricts, so that there were over 2000 registrars in all (Registrar General 1839: 269). The post of Registrar could provide a welcome supplement to the income of doctors without large private practices (Novack 1972, Checkland and Lamb 1982).

The GRO in London was to form the centre of this new network, completing the work of abstracting local information into a report to go 'docilely' before Parliament each year. In Farr's words, the living and dying realities would be:

> divested of all colour, form, character, passion and the infinite individualities of life: by abstraction they are reduced to mere units undergoing changes as purely physical as the setting stars of astronomy or the decomposing atoms of chemistry (GRO 1875: iii).

So here we have at least part of the network within which the standardised mortality ratio emerged: an unlikely combination of reformers, actuaries and impecunious Poor Law medical officers. The actuaries, as we have seen, had been producing life tables for many decades, using special surveys of individual towns. It had been a struggle, because of military and other fears, to persuade governments to undertake censuses, and a further struggle to establish the practice of collecting information on people's ages. But once ages of both living and dead were available, the actuaries' method could be brought into play for the whole population.

The length of time which elapsed before age standardisation was fully accepted shows that it was not the existence of this network, with its centre at the GRO, which alone explains the way in which the SMR has come down to us as a hard measure. William Farr was the leading medical statistician in England and Wales for two decades, and had more influence than any other individual on the development of vital statistics. In fact he made only intermittent use of standardisation, and never published age standardised mortality rates for the whole population. His major use of the method was in his debates with the 'Darwinists' in defence of sanitary reforms (Humphreys 1975: 111).

The Darwinists were a group active in the 1860s–1870s, who believed that sanitary improvement could only lead to attenuation of the principle of the survival of the fittest. If sickly children survived, the fitness of the population could be at risk. They pointed out that although sanitary engineering works proceeded apace in the mid-19th century, death rates overall did not appear to improve.[3] In the 1860s different views emerged on how to interpret those trends now made visible by censuses and death registration. One view was that if sanitation, which was expensive and a burden on the rates, was to be successful in lowering the high rates of infant mortality, the unfit would survive and become a further burden on the rates later on. On this analysis, expensive sewers would do no more than produce expensive sickly adults (Humphreys 1975). In 1861, Farr's Supplement to the 35th Annual Report admitted that while life expectancy at birth had increased, life expectancy in 'older ages' (over 35) was if anything decreasing; in the 1860s the average life expectancy in England was 49 years (GRO 1864).

Such arguments were a major challenge to the advocates of public health reform. Some opponents, such as Francis Galton, argued that because infants had an inevitably high risk of death, a high level of total population mortality was simply a sign of high birth rates. This was in turn a sign of prosperity, because when people are earning more they are more likely to marry and have children. As a fixed proportion of infants was held to be born with congenital defects, a higher number of births meant a higher number of defective babies. If these then all died, nothing was lost and perhaps there was a gain to the community from the survival of the fittest. Deaths at very early ages should therefore be removed altogether from vital statistics, as these did not indicate anything meaningful in terms of population health (GRO 1875).

The debate on how to measure life expectancy was, therefore, tied up with a debate on what constituted a 'healthy population'. Farr, and his successor William Ogle, resisted any idea that saving young lives was a waste of effort and therefore resisted removing deaths at younger ages from the vital statistics. However, they *were* concerned with age structure. As it was thought that people reproduced in the countryside and then migrated to the town to work, it was feared that the younger age structure of the rural population could lead to a misleadingly high crude death rate in country areas and a misleadingly low one in towns. For Farr and his reforming contemporaries the problem was how to persuade policy makers that the towns were unhealthy when their age structure might tend to make for a lower death rate than their 'sanitary condition' would warrant (GRO 1864).

These different sets of motives for standardisation may explain why it was that, although Farr used something very close to what we would now recognise as an SMR, he did so only now and then. For most purposes he continued to use crude death rates. For purposes of political argument he used a different type of 'standard population', that is, the so-called 'healthy districts'. And this population was not characterised by its age structure at all but by its mortality rate: 17 per thousand (Humphreys 1975: 128–30). This was the crude death rate in the healthiest districts of England and Wales (those with lowest crude mortality rates), and Farr used it as a standard in another sense; as the standard to which all districts should aim.

In the 1870s Farr's method came in for criticism from two groups: Medical Officers of Health (MOsH) from industrial cities and those from seaside resorts and spas. Despite Farr's earlier fears, it was these places (with older age structures) which tended to show up badly in the statistics that used crude mortality rates. Eyler (1976) points out that medical officers of industrial cities used the Manchester Statistical Society and the Society of MOsH to air their grievances (the Metropolitan Association of MOsH was set up in 1856). Although by no means all industrial and seaside MOsH became Darwinists, many considered the stigmatisation of their towns' high mortality rates a professional slur.

In reply, Farr constructed tables of life expectancy at different ages for what he called the 'healthy districts'. The 'Healthy District Life Table' in his section of the *33rd Annual Report of the Registrar General* was constructed in 1859 from data extending over the period 1849–1853, and took as its standard population the 63 districts of England and Wales with a death rate under 17 per 1000 per annum (Farr 1872a). The life table seems to have had several purposes. It was accompanied by calculations of 'Annuities and Premiums' and commended to its readers as having been 'found by experience' to express 'very accurately the actual duration of life among the clergy and other classes of the community living under favourable circumstances' (Farr 1872a: 441–2) (and likely to be taking out life insurance). In this case at least it seems that Farr used the healthy districts as a way of calculating life expectancy in better-off social groups. Towns and cities could also use the table to compare themselves with the 'healthy districts' regardless of their age structure.

But Farr had yet another aim. He called this 'life table method' a 'biometer' and likened it to the use of thermometers in the 'more exact' sciences. With its aid he hoped to discover the fundamental laws governing life expectancy. Such laws would in turn bestow on public health the same kind of scientific aura which surrounded the natural sciences (as reflected in his references to astronomy and chemistry cited above).

The routine use of age-standardisation by the GRO had to await Farr's successor Ogle in the Decennial Supplement for 1871–1880. It was in the annual Registrar General's Report of 1883 that standardisation to the age structure of the whole population (not to those of the 'healthy districts') first appeared as a way of correction for the varying age structures of populations. At this time there seems to be acceptance that sanitary reforms had improved population health in many of the areas where they had been strictly carried out (GRO 1885). But there also seemed to be a rising concern to exonerate sanitary authorities from blame for those urban areas where mortality was still very high (Liverpool seems to be the perpetual example of failure of health to improve). By this time no appeal was being made to Darwinist ideas on the role of unfit babies surviving; behaviour had taken over from genetics, at least temporarily (though see Eyler 1989), as the reason proffered for persisting poor health in some areas. The state of sanitation in the towns was exculpated.

Szreter (1991) has pointed out two important changes in the use of age-standardisation, and has suggested a very powerful argument for why this happened. He suggests that, under Ogle, the GRO had to address a different audience from that addressed by the visionary writings of Farr. Chadwick and Farr had wanted to reach a wide audience of reformers and concerned lay people with a single, simple method of measurement which displayed health differences in a striking manner. By the time Ogle inherited Farr's position, public health had become far more 'medicalised'. Professional and professionalising groups had occupied the territory of public health, and these were now the audience for the GRO. Ogle's audience was therefore technically more sophisticated, and used this expertise in their critiques of figures unfavourable to their interests. Szreter's argument is consistent with the theme of this paper. He has made the important point that a historically developing professional division of labour itself creates the multiplication of more specialised groups with more complex conflicting or common interests; a useful correction to the shorter time perspective of many case studies using the translation approach.

Two major critical attacks seem to have been made on the SMR. The first was made in 1922 in a special report on the measurement of life expectancy to the relatively young Medical Research Council (Brownlee 1922). By this time the notion of 'degenerative' disease had been accepted, and Brownlee was asking, are the same 'measures' in both senses, applicable to epidemic and to degenerative disease? And this led him to try and break open the black box bequeathed by Farr to Ogle, the use of standardised death rates. He appears to have been concerned to distinguish different types of aetiological factors for the *same* diseases experienced as causes of mortality by *different* age groups. Phthisis (TB) at young ages, for example, was seen as a rural disease associated with insufficient diet and sitting in wet clothing. Phthisis death in middle age on the other hand he regarded as 'associated with depression of health due to industrialism' (Brownlee 1922: 52). By this time it was no longer thought that population density and sanitary engineering were the major reasons for variations in mortality in different districts, but why were the same districts as ever found to be 'healthy' (rural) and 'unhealthy' (Liverpool etc)? Brownlee's answer was that 'the conditions which tend to depress vitality also tend to produce premature senility' (1922: 65). Once again, these were ideas which had been put forward by Neison in the 1840s (Neison 1844, 1845), but not widely accepted at that time. The report proposed a new method of measurement, the 'life table death rate', a more elaborate form of standardisation, to demonstrate the strength of these new aetiological theories. This method was never adopted.

The second major critique of the SMR was made by Yule in 1934. He read a paper to the Royal Statistical Society which showed that the SMR was unreliable for comparing populations of widely different age structures (Yule 1934). Like Neison's critique of Chadwick, however, Yule's paper is not widely remembered, and did not change practice, despite the eminence of the author. The rationale for age-standardisation did however seem to change according to the arguments put forward: these in turn were shaped by contemporary policy debates. Neison's original aim seems to have been to show that occupational hazards were more important influences on the health of the urban working classes than sanitary conditions. By the 1930s, attention had shifted firmly from regional to occupational inequalities in health. The SMR was duly accepted as *the* way to compare mortality in occupations and social classes. At the same time, various theories emerged to explain these differences, and here we begin to see the outlines now familiar in the Black Report's (Townsend and Davidson 1982) explanations: genetic selection, material factors, behaviour. But whatever explanations were being tested, studies and official statistics continued to use the SMR despite the technical characteristics of the measure which make it, for some critics, such as the American epidemiologist Mary Dempsey (Dempsey 1947) over-dependent on deaths at older ages.

According to the translation theorists (*e.g.* Latour 1987), a fact or artefact which joins together widely disparate social groups will appear to be a relatively 'hard fact'. The more socially complex and conflict-ridden the context of fact construction, the more skillfully the underlying conflicts of interest must be written-out: this is why all parties are able to agree and a claim becomes an accepted fact (or established measure). In this light, the virtue of the SMR seems to be that it developed into its present form rather slowly, being adapted at each step to shifting constellations of interests, and contributing, at each step, to a new consensus. Its persistence may then best be understood as a product of the very wide disparity of views on public health held over the period 1830–1910. In subsequent times, the distance between opposing parties in the field of public health has never been as wide. As a result, the field has seen the fleeting appearance of a raft of measures such as Years of Potential Life Lost (YPLLs), Health Life Expectancy, Quality Adjusted Life Years (QUALYs) and Disability Adjusted Life Years (DALYs). These have not been called upon to mediate the same degree of conflict as that which accompanied public health debates in the later 19th century, so that they can drift in and out of fashion in a rather indefinite manner: neither becoming 'hard measures' themselves, nor requiring to be replaced by even 'harder' ones in order to recede from attention.

The search for the cause of cancer: the advent of the odds ratio

The odds ratio is as equally firmly established in epidemiology and public health as the SMR, and it is instructive to examine its far more recent history.

Studies suggesting that smoking was a cause of lung cancer had been reported in the German literature from the late 1920s onward, with two formal investigations which would today be recognised as case-control studies appearing before 1945. One of these (Schairer and Schöniger 1943), in retrospect, appears as a methodologically sound piece of research with convincing results. This research was, however, strongly associated with the racial hygiene movement in Nazi Germany, and with the Institute for the Struggle against the Dangers of Tobacco, founded with 100,000 Reichmarks of Adolf Hitler's personal finances. The Institute disappeared at the end of the Second World War, and its legacies, including the case-control study of smoking and lung cancer, faded from view (Davey Smith, Ströbele and Egger 1994, 1995).

Since the 1920s, a rise in cancer mortality had been noted in UK official reports. By the 1950s, the observed increase in cancer mortality had come to be attributed almost entirely to lung cancer (Case 1956). Analysis of birth cohorts (*e.g.* men born around 1880, 1890, 1900 etc) revealed that lung cancer mortality rates increased to roughly the same extent for all ages within a given cohort (Kortweg 1951). This suggested that a carcinogenic agent had been introduced at some stage in the past which had affected adults at all ages from the time when it was introduced (a 'period effect'). Most commentators accepted that this was a real increase, not an artefact of changing diagnostic practices or of an ageing population. A number of environmental changes were suggested as possible causes, for example by Cutler (1955). They included increased smoking, atmospheric pollution due to car exhausts and factory emissions and increased occupational exposure to known carcinogens at the workplace.

At the outset of the debate on the causes of lung cancer, large urban-rural differences in lung cancer mortality rates seemed to support a role for atmospheric pollution. In 1947 the Medical Research Council convened a conference to consider the possible reasons for the rise in lung cancer. Cigarette smoking was only one of the possibilities put forward (Himsworth 1982, Webster 1984: 15). Stocks and Campbell concluded that 'about half the deaths of Liverpool men from lung cancer arise from cigarette smoking and about three-quarters of the remaining half were due to a factor which is only slightly present in the rural area'. They considered that this factor was likely to be 3:4-benzpyrene, other polycyclic hydrocarbons or sulphur dioxide in the atmosphere (Stocks and Campbell 1955: 929).

In 1951 Jerome Cornfield of the Office of Biometry of the National Institutes of Health (a major 'center of calculation' in the US health establishment) acknowledged that:

> A frequent problem in epidemiological research is the attempt to determine whether the probability of having or incurring a stated disease, such as cancer of the lung, during a specified interval of time, is related to the possession of certain characteristics, such as smoking (Cornfield 1951: 1270).

Some researchers were willing to consider that both smoking and atmospheric pollution were causes of lung cancer. Others considered that smoking might not be causally involved at all. The eminent statistician R.A. Fisher, for example, suggested that a constitutional factor could be related both to a tendency to smoke and to an increased risk of lung cancer (Fisher 1958), producing an association that was spurious. Hans Eysenck, like Fisher working with the support of the tobacco industry, produced data which he claimed showed that the personality profile of smokers was different from that of non-smokers (Eysenck *et al.* 1960), and that personality could predispose to cancer.

The critics of the smoking hypothesis referred to the epidemiological data linking smoking with lung cancer as 'just statistics'. This was a reference to the absence of truly 'experimental' studies of the link. Cancer sufferers were distinct from non-sufferers in a large number of ways. How could 'true causes' be distinguished from accidental associations or 'confounders' under these circumstances? Such statements invoked evident annoyance in Cornfield. For him and his colleagues '[t]he differentiation between various methods of scientific enquiry escapes us as being a valid basis for the acceptance or rejection of facts' (Cornfield *et al.* 1959: 202).

One of the pioneers of the aetiological investigation of lung cancer, Austin Bradford Hill, on the other hand, accepted that:

> It is, of course, possible that the relative absence of nonsmokers and the relative frequency of heavy smokers that Doll and I found in our patients with cancer of the lung is really a function of some other difference between the two groups. We do not ourselves . . . believe this is so . . . Nevertheless here lies, I admit, the weakness of the observational data as compared with the experimental approach (Hill 1953: 8).

The problem facing the epidemiologists, who needed to enrol the same allies (including clinicians and funding agencies) as the experimental biologists and other types of scientist involved in the study of cancer, was how to demonstrate the relative importance of different potential 'causes'. The experimental method had high status as the classical method for scientific work. Lung cancer (despite its status as a dread disease) is a relatively rare condition. Most accounts of its aetiology regard it as the outcome of exposure to a causal agent over a period of time, possibly followed by another long period of latency. It is therefore not amenable to the same type of experimentation as, for example, infectious diseases which are produced quite rapidly by exposure to a causal agent. In lung cancer and other chronic diseases epidemiologists could only carry out experiments with great difficulty.[4]

The decisive work proving that smoking is the cause of lung cancer was carried out by Doll and Hill (1950).[5] Their data consisted of a comparison of the prevalence of risk factors among 'cases' – patients with lung cancer – and 'controls' without the disease. There had been several previous studies of lung cancer which also showed an excess of smoking in these patients (for example Muller 1939, Schrek *et al.* 1950), but these had convinced no-one. One technical problem in this new type of study (the 'case-control study') was that people could not be randomly allocated into exposed and non-exposed groups. There was always the possibility therefore that smokers and non-smokers could have selected themselves, and could therefore be different in other ways possibly relevant to the causation of cancer.

The inability to randomise was of course a more general problem for those who wished to claim for the relatively new discipline of epidemiology a truly scientific status on a par with experimental sciences. It is generally difficult, even impossible, to experiment on populations. Cornfield set out to try and remedy this. It had to be admitted that those who smoked might differ in some other way (personality for example) from those who did not. In the paper quoted above (Cornfield *et al.* 1959), he therefore proposed several summary measures of mortality risk which could be calculated from observational data. One of these methods of measurement was an estimator of the relative risk of disease (*e.g.* lung cancer) in the exposed group (*e.g.* in smokers) divided by the risk of the disease in the unexposed group (non-smokers). This was later to be christened the 'odds ratio' although Cornfield did not call it that until later.

He suggested that the advantage of this method was that it could be interpreted as a measure of the strength of an association. The groups with and without disease might differ in several ways, so that more than one factor might appear to be associated with lung cancer. But Cornfield argued that the most important aetiological factor was that with the *largest* relative risk. The bigger the odds ratio the stronger the link between the proposed causal factor and the disease. The method was quickly and widely accepted and used by other scientists (Dorn 1959), and became regarded as the 'natural' way to analyse data from observational epidemiological studies. It could serve a range of interests: epidemiologists in their claims for scientific status; as well as a number of industries other than the tobacco companies whose products had fallen under suspicion as possible contributors to the rapid increase in cancers (Proctor 1995).

Once accepted as *the* correct method for measuring association, a simple transformation of the role of the odds ratio could be performed, constituting it as a way to decide on the relative importance of competing potential causes of disease (Cornfield *et al.* 1959: note 22). There were many factors: atmospheric pollution, 'constitution', personality – 'exposure' to which might be the same in cases and controls. All of these had been advanced as competitors of smoking as the true cause of lung cancer. But from now on these would have to be shown to have a stronger relationship with the disease in the sense of having a larger odds ratio, or be discarded.

Once this rule (referred to as Cornfield's Law) was accepted, atmospheric pollution (for example) could no longer be taken seriously as a cause of lung cancer, since the odds ratio for pollution was considerably smaller than that for smoking. As Cornfield and colleagues wrote:

> The magnitude of the excess lung cancer risk among cigarette smokers is so great that the results can not be interpreted as arising from an indirect association of cigarette smoking with some other agent or characteristic, since this hypothetical agent would have to be at least as strongly associated with lung cancer as cigarette use; no such agent has been found or suggested (Cornfield *et al.* 1959: 202).

And any other proposed aetiological agent would from then on have to pass the test of having the same or greater association according to this method. We now take this idea so much for granted that our reaction may well be 'so what'? Fisher and Eysenck are regarded as having mysteriously blotted their scientific copybooks by taking an opposing view, just as John Simon and Farr opposed Snow's water-borne theory of cholera and Koch proposed to use tuberculin to inoculate people with active TB (Vandenbrouke 1989). Stolley employs an implicit 'interest model' to explain the deviant behaviour of Fisher in questioning the smoking-cancer link, pointing out that Fisher was at this time at the end of his career. He had ideological objections to mass public health campaigns. He was not good with large data sets. He was at a loose end in retirement and craved the excitement of controversy. Some even argued that he had 'taken a fee from the tobacco industry, although those who know him best doubted that the fee mattered very much' (Stolley 1991). Those who opposed the account of risk which we now take as the correct one were therefore attributed 'interests' and 'motives'. Vandenbrouke points out that the technical characteristics of their criticisms of the odds ratio do not, however, differ from critiques of 'poor method' taught to the students of today. He regards these critiques as:

> extremely well-written and cogent papers that might have become textbook classics for their impeccable logic and clear exposition of data and argument if only the authors had been on the right side (Vandenbrouke 1989: 3).

After all, all that Fisher was doing was applying the text book warning against mistaking correlation for causation, and making a point which might nowadays prove popular about the neglect of genetic factors by epidemiology and public health (Himsworth 1984, Strong 1990). Like schizophrenia in the 1990s, smoking had recently been shown to be more concordant in monozygotic than dizygotic twins. Eysenck (Eysenck *et al.* 1960) and Fisher (1958) speculated on underlying genetic factors predisposing to both smoking and to lung cancer (later studies of smoking in twins provided contrary evidence to this (*e.g.* Braun *et al.* 1994)).

Vandenbrouke points out that the other critic of Doll and Hill, Berkson, was also making a textbook point about the superiority of randomised control trials to case control studies in determining the differential risk of disease (Vandenbrouke 1991). Vandenbrouke concludes:

> If . . . one would replace the words 'cigarette smoking' and 'lung cancer' in these papers by 'Exposure X' and 'Disease Y', one could obtain extremely clear exposés of confounding and some other intractable issues . . . In themselves, the [critical] papers look so convincing that I wonder whether I would have belonged to the sceptics if this were the 1950s or early 1960s,

and that:

> The viability of a thought might well be dictated by convenience or extra scientific influence (Vandenbrouke 1989: 5).

In his later paper (Vandenbrouke 1991) he reveals that Doll and Hill's early work is still used to demonstrate what can go wrong in case control studies. Previous papers had failed to swing medical opinion. Why therefore was it influenced by Cornfield's odds ratio? Wolff (1992) points out that, in his opinion, the earlier, ignored papers had used a technically superior method, and that the acceptance of Doll and Hill's findings was untypically rapid. He postulates two more actors in the 'agonistic field' of the debate on the causes of cancer: the motor vehicle and oil industries. The tobacco theory and its associated method of measurement won out because its allies were stronger ones.

Why was it that a given set of statements about lung cancer came to be accepted by more or less all the important players in oncology, epidemiology and public health (as well as a number of other less obvious players)? In the terms of the translation programme, one interpretation of the statistical picture presented by lung cancer became an 'obligatory point of passage' (Latour 1987). Each of the groups involved in the field could gain more by agreeing to it than by rejecting it, perhaps for very disparate reasons. Wolff's (1992) work suggests the possibility of an alliance between epidemiologists seeking to advance their discipline, the UK Medical Research Council (where both Brownlee and Bradford Hill had positions), the US National Institutes for Health (Cornfield's employers at one point) preserving their institutional power, and certain industrial groups. This is by no means the only possibility, and we need to remember that it is the very distance between those groups united in their construction of hard facts and measures that gives them their resilience.

Conclusion

The future task for a sociology of measurement in health is great. Not least would be the task of delving behind the published work into the archives of the 'centres of calculation' and individual research units involved, the Departments of Health, and funding bodies of the research such as the Medical Research Council, the Nuffield Foundation. In this chapter we have done no more than suggest a methodology and begun to illustrate some of its uses. We have not followed the full intricacy of the alliances that produced the SMR or the odds ratio, but only pointed in some directions.

Far more controversy and network-building was involved in superseding the 'average age at death', and in establishing the SMR and the odds ratio, than we had suspected at the outset of this study. The present chapter does no more than point towards a rich field for sociological and historical studies of medical measurement. Our case studies have used a method of analysis which concentrates not only on the inventors of methods of measurement, but also on their allies and even their antagonists. In each case the network of fact-builders was a wide one. It would also have been misleading to take up a position based on what we now 'know' to be 'the right method' and expect that, or the nearest offer, to be accepted without controversy or to lead to policies of reform. The Chadwick/Neison debate had no effect because Neison's critique of the average age at death was ignored, as were Fisher's, Eysenck's and Berkson's ideas on smoking and lung cancer. At a much later date, however, Neison's method of age standardisation *was* to be accepted, as was his emphasis on occupation rather than region as the most significant site of mortality difference. Yule's criticisms of the SMR might be considered just as intellectually cogent as Fisher's of the odds ratio, with equally little effect.

What the case of the odds ratio shows is that a method may be justified by the acceptance of the representation of nature it provides, as much as by its abstract logic: rather like the rules of a game which allow players, even as antagonists, to arrive at a shared objective: a mutually agreed outcome and the continuation of the game itself. What the analytical observer needs to discover is why all parties have agreed on the same set of rules. In order to do this we need to identify the parties and observe the negotiations involved.

In all three case studies we have seen fact-builders forming alliances and striving to extend and strengthen the networks leading out and back from 'centres of calculation'. Indeed one of the most obvious differences between 'those who were wrong' and those whose methods of measurement were accepted was the relationship of the 'victors' with state policy-making processes and organisations. Neither Neison nor Yule had the same sorts of links to policy-making bodies as Chadwick or Farr, and in the later debates both Brownlee and Bradford Hill worked at some point for the British Medical Research Council,[6] and Cornfield for the American National Institutes for Health.

A lone voice, offering a new method of measurement, is not sufficient: the new technique must be taken up by influential groups as relevant to their interests. This paper has only begun to sketch the intricate strategies of scientists such as Chadwick, Neison, Farr and Cornfield. They and their allies and opponents acted within a context of political and commercial institutions. 'Centres of calculation' such as insurance companies, General Register Offices and National Institutes for Health do not just appear, someone has to pay for them. A national Census and a GRO for England and Wales were made possible by the alliance between groups with diverse interests. Dissenters insisted that registration should not be done by Anglican parishes; actuaries insisted that Censuses record age; impecunious doctors agitated for a monopoly over death registration. Once established, 'centres of calculation' struggled to preserve their claim to resources and Cornfield's efforts were part of such a struggle. We can see parallels today when governments threaten to rationalise or close down the Office of National Statistics (until 1997 the Office of Population Censuses and Surveys, the successor to the GRO) and other sections of the government statistical service. One response to this threat has been a debate on whether to widen possible alliances by including behavioural factors such as smoking as well as material ones such as class and housing tenure in Censuses (Bartley 1990). Another has been to make Censuses more relevant to health service planning by including a question on morbidity in the 1991 Census (once again, something for which Farr agitated without success).

These shifting alliances are important in two ways: first, because they may result in the inclusion of new items (such as age and morbidity) in routine statistics which make it possible to build new kinds of facts; secondly, because they appear to have considerable influence on the methods adopted to combine these items into facts about levels of risk. At various times an estimate of 'true life expectancy' or 'true risk of death' is reached which is good enough for the practical purposes of most of the groups involved. When this happens, it is solidified into a 'black box'. Estimates are accepted as 'accurate' when they become a point of passage through which groups such as insurance societies, government actuaries, dissenters, epidemiologists and public health reformers and planners can reach their several objectives. It is the firmness of the alliance which creates the 'truth' of the fact or the correctness of the method of measurement. A firm alliance can get away with leaving out some influential groups. In such cases the knowledge claims of the excluded will be disqualified from 'rational debate'. For example, the case-control design may be seen as no more than a form of machination by political extremists; or the scientists whose work accords with the interests of the tobacco companies may be branded as uniquely 'motivated by vested interests' and subject to irrational error, or even as mentally unbalanced.

As these alliances typically include actors within the policy making arena as well as scientists, we also need to take a sceptical attitude to the notion of 'the impact of research on policy'. Perhaps true facts and correct methods of measurement are as much an outcome of the policy process as a possible influence upon it?

Our case studies have also shown that it is not just the final form of a method of measurement which needs to be examined, but the steps by which each item becomes possible (civil registration, the inclusion of age in the Census, centralisation of death and census data in one organisation). Such an analysis has considerable implications at a time when new methods of measurement are being sought to guide health policy. The conditions under which medical and health services research are carried out may have a major effect upon the type of measures which are developed for future use. Different alliances may construct different pictures of aetiology and prevalence. If the nature of these alliances is made more explicit, some apparent contradictions may be clarified.

Appendix: The Standardised Mortality Ratio

We realise that not everyone interested in the general topic of this paper may be familiar with the way in which the SMR is, at present, calculated. The principle is simple. To obtain the SMR for persons aged 15 to 65:

1. Divide the population aged 15 to 65 of, *e.g.* England and Wales into ten-year age groups, separately for men and women.
2. Get a mortality rate for each of these groups.
3. Divide the group you are interested in (*e.g.* a social class, say class IIIn) into the same age and sex groups.
4. Apply the rates in each of the age groups in step 2 to the numbers of each age/sex group in the social class you are interested in. So for example, if the population

death rate for 45–55-year-old women in step 2 is 10 per cent and you have 200 people that age in social class IIIn then this gives you 20 deaths *expected* in that class at ages 45–55.

5. Add up all the *expected* deaths in each age group, to get the expected number of deaths in the social class as a whole. Say this was 12 deaths in each age group, with five age groups between 15 and 65, this would give 60 expected deaths.
6. See how many women in class IIIn actually died during the period of interest. This gives you the *observed* number of deaths. Say it was 80.
7. The SMR is *observed/expected* × 100. Here this would be 80/60 × 100 = 133.33. The SMR for class IIIn women is 133. The SMR summarises the extent to which the observed number of deaths in a group is greater (or less) than would be expected if the age composition of the group were identical with that of the population as a whole.

Notes

1 See Appendix for an explanation of the SMR.

2 This is presumably a reference to the Chartist struggles. The state's response to these, by enlarging the franchise in 1834, had led, according to Nissel, to the debates which resulted in the Registration Bills in 1835 which were enacted in 1836.

3 Not all followers of the ideas of Darwin opposed public health measures, however.

4 It may be noted here that in the 1980s, Rose *et al.*'s (1982) randomised control trial – a 'true experiment' – of anti-smoking advice did not demonstrate that giving up smoking led to higher life expectancy in the group that quit, although the rate of lung cancer was somewhat reduced. We should perhaps be careful not to fall into an easy acceptance of the 'difficulty of the experimental method' but be aware that this, too, is constructed within networks of influence.

5 Interestingly, it is only British-trained epidemiologists who regard Doll and Hill's work as 'decisive': in a rather similar phenomenon to the chauvinistic lionisation of Pasteur in France described by Latour (1984, 1987). Text books in the United States attribute the same finding to two papers which appeared in the *Journal of the American Medical Association* several weeks before Doll and Hill's (1950) paper appeared in the *British Medical Journal*.

6 Bradford Hill's father was head of the Department of Applied Physiology at the early Medical Research Council, and Hill's first post was at the Industrial Fatigue Research Board, a semi-independent body under the auspices of the MRC (see Hill 1982).

References

Armstrong, D. (1986) The invention of infant mortality, *Sociology of Health and Illness*, 8, 211–32.

Bartley, M. (1985) 'Coronary' heart disease and the public health 1950–1983, *Sociology of Health and Illness*, 7, 289–313.

Bartley, M. (1990) The story of r_2. In Varcoe, I., McNeil, M. and Yearley, S. (eds) *Deciphering Science and Technology*. London: Macmillan.
Blane, D.B., Davey Smith, G. and Bartley, M.J. (1990) Social class difference in years of potential life lost: size, trends, and principal causes, *British Medical Journal*, 301, 429–32.
Braun, N.J., Caporaso, N.E., Page, W.F. and Hoover, R.N. (1994) Genetic component of lung cancer: cohort study of twins, *Lancet*, 344, 440–3.
Brownlee, J. (1922) The use of death rates as a measure of hygienic conditions, *Medical Research Council Special Reports Series, no. 60*. London: HMSO.
Case, R.A.M. (1956) Cohort analysis of mortality in England and Wales 1911–1954 by site and sex, *British Journal of Preventive and Social Medicine*, 10, 172–99.
Chadwick, W. (1844) On the best modes of representing accurately, by statistical returns, the duration of life and the pressure and progress of the causes of mortality, *Quarterly Journal of the Statistical Society of London*, 7, 1–40.
Charlton, J., Wallace, M. and White, I. (1994) Long term illness: results from the 1991 Census, *Population Trends*, 75, 18–25.
Checkland, O. and Lamb, M. (eds) (1982) *Health Care as Social History*. Aberdeen: Aberdeen University Press.
Collins, H.M. (1985) *Changing Order*. London: Sage.
Cornfield, J. (1951) A method of estimating comparative rates from clinical data. Applications to cancer of the lung, breast and cervix, *Journal of the National Cancer Institute*, 11, 1269–75.
Cornfield, J., Haenszel, W., Hammond, E.C., Lilienfeld, A.M., Shimkin, M.B. and Wynder, E.L. (1959) Smoking and lung cancer: recent evidence and a discussion of some questions, *Journal of the National Cancer Institute*, 22, 173–203.
Coutouzis, M. and Latour, B. (1986) The solar village of Frangco-Castello, *L'Année Sociologique*, 36, 114–66.
Cullen, M.J. (1975) *The Statistical Movement in Early Victorian Britain*. New York: Hassocks.
Cutler, S.J. (1955) A review of the statistical evidence on the association between smoking and lung cancer, *Journal of the American Statistical Association*, 50, 267–82.
Davey Smith, G., Ströbele, S.A. and Egger, M. (1994) Smoking and health promotion in Nazi Germany, *Journal of Epidemiology and Community Health*, 48, 220–3.
Davey Smith, G. Ströbele, S. and Egger, M. (1995) Smoking and death (letter), *British Medical Journal*, 310, 96.
Dempsey, M. (1947) Decline in tuberculosis: the death rate fails to tell the entire story, *American Review of Tuberculosis*, 56, 157–64.
Doll, R. and Hill, A.B. (1950) Smoking and carcinoma of the lung: preliminary report, *British Medical Journal*, 2, 739–48.
Dorn, H.F. (1959) Some problems arising in prospective and retrospective studies of the etiology of disease, *New England Journal of Medicine*, 261, 571–9.
Drever, F. and Whitehead, M. (1995) Mortality in regions and local authority districts in the 1990s: exploring the relationship with deprivation, *Population Trends*, 82, 19-26.
Eyler, J.M. (1976) Mortality statistics and Victorian health policy: program and criticism, *Bulletin of the History of Medicine*, 50, 335–55.

Eyler, J.M. (1989) Poverty, disease, responsibility: Arthur Newsholme and the Public Health Dilemma of British Liberalism, *Millbank Quarterly*, 67-Suppl I, 109–26.

Eysenck, H.J., Tarrant, M. and Woolf, M. (1960) Smoking and personality, *British Medical Journal*, i, 1456–60.

Farr, W. (1872a) *Appendix to the 33rd Annual Report of the Registrar-General of England and Wales*. London: Eyre and Spottiswode.

Farr, W. (1872b) Inaugural Address, *Journal of the Statistical Society of London*, 35, 417–30.

Farr, W. (1837–8) On benevolent funds and life assurance in health and sickness, *Lancet*, i, 701–4 and 817–23.

Fisher, R.A. (1958) Lung cancer and cigarettes? *Nature*, 182, 108.

General Register Office (1847) *Annual Report of the Registrar General for England*. London: HMSO.

General Register Office (1864) *25th Annual Report of the Registrar General of Births, Marriages and Deaths in England*. London: HMSO.

General Register Office (1875) *Supplement to the 35th Annual Report of the Registrar General for England: Mortality 1861–1870*. London: HMSO.

General Register Office (1885) *Supplement to the 45th Annual Report of the Registrar General of Births, Marriages and Deaths in England*. London: HMSO.

Goldman, L. (1991) Statistics and the science of society in early Victorian Britain: an intellectual context for the General Register Office, *Social History of Medicine*, 4, 415–34.

Greenwood, M. (1946) Medical statistics from Graunt to Farr, *Biometrika*, 33, 1–24.

Guy, W.A. (1845) On the duration of life among the families of the peerage and baronetage of the United Kingdom, *Journal of the Royal Statistical Society of London*, 8, 69–76.

Guy, W.A. (1846) On the duration of life among the English gentry, with additional observations on the duration of life among the aristocracy, *Journal of the Royal Statistical Society of London*, 9, 37–49.

Higgs, E. (1991) Disease, febrile poisons and statistics: the census as a medical survey 1841–1911, *Social History of Medicine*, 4, 465–78.

Hill, A.B. (1953) Observation and experiment, *New England Journal of Medicine*, 248, 3–9.

Hill, A.D. (1982) Austin Bradford Hill – Ancestry and early life, *Statistics in Medicine*, 1, 297–300.

Himsworth, H. (1982) Bradford Hill and statistics in medicine, *Statistics in Medicine*, 1, 301–3.

Himsworth, H. (1984) Epidemiology, genetics and sociology, *Journal of Biosocial Science*, 16, 159–76.

Humphries, N. (ed) (1975) *William Farr 1807–1883. Vital statistics: a memorial volume from the reports and writings of William Farr*, 2nd Edition. New Jersey: Metuchen.

Kortweg, R. (1951) The age curve in lung cancer, *British Journal of Cancer*, 5, 21–7.

Latour, B. (1984) *Les Microbes: Guerre et Paix*. Paris: Editions Metailie, Collection Pandore. (Re-published in English as *The Pasteurization of French Society*. Harvard University Press 1989.)

Latour, B. (1987) *Science in Action*. Milton Keynes: Open University Press.

Latour, B. (1988) A relativistic account of Einstein's relativity, *Social Studies of Science*, 18, 3–44.

Macleod, R. (ed) (1988) *Government and Expertise*. Cambridge: Cambridge University Press.

Muller, F.H. (1939) Tabakmissbrauch und Lungencarcinom [Tobacco misuse and lung cancer], *Zeitschrift für Krebsforschung*, 49, 57–85.

Neison, F.G.P. (1844) On a method recently proposed for conducting inquiries into the comparative sanitary conditions of various districts, with illustrations, derived from numerous places in Great Britain at the period of the last census, *Journal of the Statistical Society of London*, 7, 40–68.

Neison, F.G.P. (1845) Contributions of vital statistics especially designed to elucidate the rate of mortality, the laws of sickness and the influences of trade and locality on health, derived from an extensive collection of original data, supplied by Friendly Societies, and proving their too frequent instability, *Quarterly Journal of the Statistical Society of London*, 8, 290–343; continued in *Journal of the Statistical Society of London* (1846) 9, 50–76.

Nissel, M. (1987) *People Count: A History of the General Register Office*. London: HMSO.

Novak, S.J. (1972) Professionalism and bureaucracy: English doctors and the Victorian public health administration, *Journal of Social History*, 6, 440–64.

Prior, L. (1985a) The good, the bad and the unnatural: some aspects of coroners' decisions in Northern Ireland, *Sociological Review*, 33, 64–90.

Prior, L. (1985b) The social production of mortality statistics, *Sociology of Health and Illness*, 7, 167–90.

Proctor, R. (1995) *Cancer Wars*. London: Basic Books.

Registrar-General for England (1839) The first annual report of the Registrar-General on births, deaths and marriages in England in 1837–8, *Journal of the Royal Statistical Society of London*, 2, 269–74.

Roberts, D. (1959) Jeremy Bentham and the Victorian administrative state, *Victorian Studies*, 2, 193–210.

Rose, G., Hamilton, P.J.S., Colwell, L. and Shipley, M.J. (1982) A randomised control trial of anti-smoking advice: 10 year results, *Journal of Epidemiology and Community Health*, 36, 102–8.

Schairer, E. and Schöniger, E. (1943) Lungenkrebs und Tabakverbrauch, *Z. Krebsforsch*, 54, 261–9.

Schrek, R., Baker, L.A. and Ballard, G.P. (1950) Tobacco as an etiologic factor in disease. I. Cancer, *Cancer Research*, 10, 49-58.

Stocks, P. and Campbell, J.M. (1955) Lung cancer death rates among non-smokers and pipe and cigarette smokers, *British Medical Journal*, ii, 923–9.

Stolley, P.D. (1991) When genius errs: R.A. Fisher and the lung cancer controversy, *American Journal of Epidemiology*, 133, 416–25.

Strong, P.M. (1990) Black on class and mortality: theory, method and history, *Journal of Public Health Medicine*, 12, 168–80.

Szreter, S. (1991) The GRO and the public health movement in Britain 1837–1914, *Social History of Medicine*, 4, 435–64.

Thevenot, L. (1987) *Forme Statistique et Lien Politique*. Paris: Institut National de la Statistique et des Etudes Economiques, INSEE note n° 112/930.

Townsend, P. and Davidson, N. (1982) *Inequalities in Health: the Black Report*. Harmondsworth: Penguin.

Vandenbrouke, J.P. (1989) Those who were wrong, *American Journal of Epidemiology*, 130, 3–5.

Vandenbrouke, J.P. (1991) Invited commentary: how much retropsychology? *American Journal of Epidemiology*, 133, 426–7.

Webster, C. (1984) *Health: Historical Issues*. London: Centre for Economic Policy Research Discussion Paper No. 5.

Wolff, S. (1992) Re: 'Invited commentary – How much retropsychology', *American Journal of Epidemiolgy*, 136, 1314–15.

Yule, G.U. (1934) On some points relating to vital statistics, more especially statistics of occupational mortality, *Journal of the Royal Statistical Society*, XCVII, 1–82.

6. The science and politics of medicines regulation

John Abraham

Introduction

With some notable exceptions, such as Davis (1996), Elston (1994) and Gabe and Bury (1996), few writers in medical sociology have concerned themselves with medical science or scientists working in medical fields other than prescribing doctors. Curiously, research in medical sociology is not often concerned with medicines themselves. This may be due to the apparently limited cross-fertilisation between the fields of medical sociology and sociology of science. Like many of the other contributors to this monograph, I aim to offer some advancement in this respect.

In an innovative and refreshing way, Bartley (1990) attempted to relate these two fields, but the extent to which medical sociologists have taken up her recommendation that there are lessons to be learned from the sociology of science seems rather minimal. Furthermore, while Bartley's proposed 'marriage' of medical sociology and sociology of science is to be welcomed in broad terms, it is rather selective and incomplete. She refers to the positivistic Mertonian sociology of science as 'the weak programme' because its account of the organisation of the 'scientific community' tends to neglect the sociology of scientific *knowledge* (SSK). According to Bartley, that programme has been overtaken by the relativistic 'strong programme' in SSK and social constructionism advocated by writers, such as Bloor (1973), Barnes and Shapin (1979) and Latour and Woolgar (1986). For Bartley, the advantage of the 'strong programme' is that it examines how the commitments of scientists to particular technical approaches, and to professional interests, construct knowledge itself. Thus, the 'modern' sociological study of scientific knowledge is seen as a liberatory path away from a scientistic view of medicine as non-sociological, and 'away from social and political interests and towards the study of technical and professional interests' (Bartley 1990: 380).

Leaving aside the many problems associated with the relativism of such SSK, which have been discussed elsewhere (Abraham 1995: 4–15, Bunge 1991, Millstone 1978), Bartley's account of the sociology of science unfortunately does not challenge the idea that the political economy of medicines, involving the interests of the State and the pharmaceutical industry, may be bypassed in favour of concerns about professional interests in technical dominance. Rather, it tends to support an imbalance in medical sociology, namely, that of attending to medicines' users rather than drugs' producers and regulators. Frequently medical sociology concerns itself with lay persons' and/or doctors' perceptions or interpretations of medicines, rather than with the production of medicines by industry or the control by the state over which drugs get to be used in the first place. The knowledge about medicine constructed by lay groups, patients, nurses, general practitioners and the mass media all seem to be of great interest to medical sociologists (and rightly so), but much less attention is given to the medical knowledge produced by the pharmaceutical industry or governments.

Bartley's conclusion, however, is by no means the inevitable one to be drawn from a review of SSK. She overlooked the self-confessed 'weak programme' in the SSK proposed by Chubin and Restivo (1983). That programme, based on ontological realism and deliberately so named in opposition to the 'strong programme', advocated the examination of the relationship between scientific knowledge and political interests and institutions, especially governments and industries. Realism appreciates, and presupposes, that scientific knowledge is produced by a combination of social organisation, including structural interests, and the natural effects of mind-independent generative mechanisms in nature (Bhaskar 1979, 1986). It further presupposes that interests are 'real' in the sense that they *may* exist independently of actors' knowledge of them, even though interests require social/societal creation. It follows from this that, in analysing scientific knowledge, medical sociology should go beyond social constructions of negotiation processes to take account of the cognitive structures and real interests of science. Realism also goes beyond Merton because it recognises the importance of scientists and the production of scientific knowledge outside academia. Hence, there is an important strand within SSK that, if synthesised with the conventional concerns of medical sociology, could broaden and deepen the field to take more account of the political and economic factors in the production of medical science. For example, medical scientists, who work for the pharmaceutical industry, regulatory authorities or governmental advisory committees, are all part of the wider profession of medicine and they impact upon the provision of medicines to the patient population.

A major purpose of this chapter, then, is to redress the balance by examining the regulation of the pharmaceutical industry with specific reference to the scientific testing of medicines and the implications of regulatory developments for consumer and industrial interests. This involves entering the jungle of 'regulatory science' (Jasanoff 1990). Within such science the expression of validated knowledge through peer review and published papers is much less important than in academic 'research science'. Much of regulatory science is not submitted to the discipline of anonymous peer review and/or publication. Nevertheless, regulatory authorities concerned with drug testing do employ highly trained medical scientists, who review the medical research conducted by industry in support of new drug products. Moreover, in technocratic style, regulators frequently seek advice from expert medical scientists who sit on advisory committees. Superficially, this process has some parallels with the editors of medical journals seeking advice from academic referees. On the other hand, as Jasanoff points out, in regulatory science, 'expert referees may either be formally affiliated with particular interest groups or otherwise have a stake in the outcome of the regulatory process' (1990: 81).

More generally, medicines regulation and its underpinning science are subjected to *interest representation*. A common conceptual and empirical theme is the extent to which regulators are vulnerable to 'industrial capture' – a term that describes a situation in which regulatory authorities are supposed to be defenders of the 'public interest' but, in practice, come to prioritise the interests of the regulated industry over the 'public interest' (Mitnick 1980). In particular, government regulation of the pharmaceutical industry is characterised by varying degrees of 'corporatism', that is, 'the granting of semi-official status to specific interest groups, which then assist the government to implement those policies that directly affect them' (Vogel 1986: 273). Traditionally, the nation-state has been the unit of analysis for corporatist theory (Cawson 1985) but, as Greenwood and Ronit (1991) have suggested, it is also important to examine the formulation of regulatory relationships at the *transnational* level. This is particularly pertinent to the study of regulatory science in the pharmaceutical sector because both the industry and its science are transnational.

This chapter combines international comparative analysis with a sensitivity to the importance of transnational developments. The focus is on Western industrialised countries because that is where the scientific research-based pharmaceutical industry and associated regulatory science are most active and developed. Arguably the direction of the global pharmaceutical industry will be determined by the industrial strategies and regulatory policies established in those countries. British regulation of chemical and technological risk is often considered by scholars to pursue a cooperative policy style with industry compared with a more adversarial relationship between industry and regulators in the US (Abraham and Millstone 1989, Brickman *et al.* 1985, Gillespie *et al.* 1979, Irwin 1985). Following this tradition the UK and the US have been chosen for particular attention in this chapter because they provide a valuable and relatively accessible framework with which to identify key sociological/political aspects of *medicines* regulation. Furthermore, that framework is helpful in analysing other regulatory systems, such as the European Union (EU), which is of interest because it is the largest single example of transnational development in the sector. Nevertheless, it is important to appreciate that regulatory science is taking on global dimensions within which the EU is only one major actor. For this reason, I also discuss some important features of the globalisation of medicines regulation.

After outlining the nature of the data sources and methodology for the research in this chapter, I explain the nature of the scientific uncertainty that underpins medicines regulation and then argue that British medicines regulation has been characterised by a corporate bias prioritising industrial interests over consumers' interests. My account of US medicines regulation acknowledges that there have been attempts at industrial capture, but these have been resisted to some degree and, compared with the British system, there is much less corporate bias in favour of industry. I then argue that the Europeanisation and globalisation of medicines control is adopting a British rather than American approach to medicines regulation and I discuss the implications of this for the medical science of drug testing and regulatory institutions and consumers' interests. Of course, it is inevitable that this approach cannot adequately examine the technical and regulatory inconsistencies *within* countries as occurred with *Depo-Provera* and *Halcion* in the UK, but these have been discussed elsewhere (Abraham and Sheppard 1995, Bunkle 1993).

Data sources and methodology

The defining feature of the research methodology was the collection of extensive documentary data supplemented by interviews. This was not necessarily a linear process. For example, interviews often led to the discovery of important documents. In fact, most of the sources for this chapter are based on documentary research. The main British documentary sources were: the trade press, such as the *Pharmaceutical Journal*, *Scrip* and the publications of the Association of the British Pharmaceutical Industry (ABPI); appropriate medical/scientific literature accessed via MEDLINE and DIALOG computer databases, especially publications by the Centre for Medicines Research (CMR); official documents produced by the Department of Health's Medicines Control Agency (MCA), its predecessor the Medicines Division and its expert advisory Committee on the Safety of Medicines (CSM). Major American and European sources included: US Congressional investigations into the operations of the Food and Drug Administration (FDA) and the pharmaceutical industry; the official publications of the European Commission on drug regulation; and the appropriate medical/scientific literature emanating from these organisations. I have also made use of publications by the World Health Organisation (WHO).

Interviews were conducted with regulators and industrialists, including medical scientists, in the UK, the US and the EU. These interviews spanned the period 1988 to 1994, including eight in the US, nine in the UK, five in Sweden, five in Germany, two in the Netherlands and three at the 'EU level'. Interviewees were 'informants', rather than 'respondents' because they were specifically selected on the grounds of their strategic positions and knowledge about the pharmaceutical industry and medicines regulation, rather than as a representative survey sample of industrialists or regulators. Nevertheless, the individuals involved were key actors in the field so their views may be seen as indicative of the leading thinking in industry and regulatory authorities. No-one refused to be interviewed, but some informants did not wish to comment on matters which they regarded as confidential. The interviews were semi-structured, involving a 'core' interview schedule, but with appropriate flexibility to adapt to the responses provided by the informants.

Regulatory science and technical uncertainty

In deciding whether or not patients should be exposed to a newly developed drug it is necessary to make some risk-benefit assessment of the prospective medicine in question. To do this medicines regulators and others turn to the scientific testing of drugs which comprises preclinical pharmacology (including efficacy, metabolic and toxicological assessments in animals), safety studies in healthy human volunteers, clinical trials with patients and a form of epidemiology, known as pharmacovigilance, which tracks patients' adverse experiences with drugs on the market (see Figure 1). Indeed, these

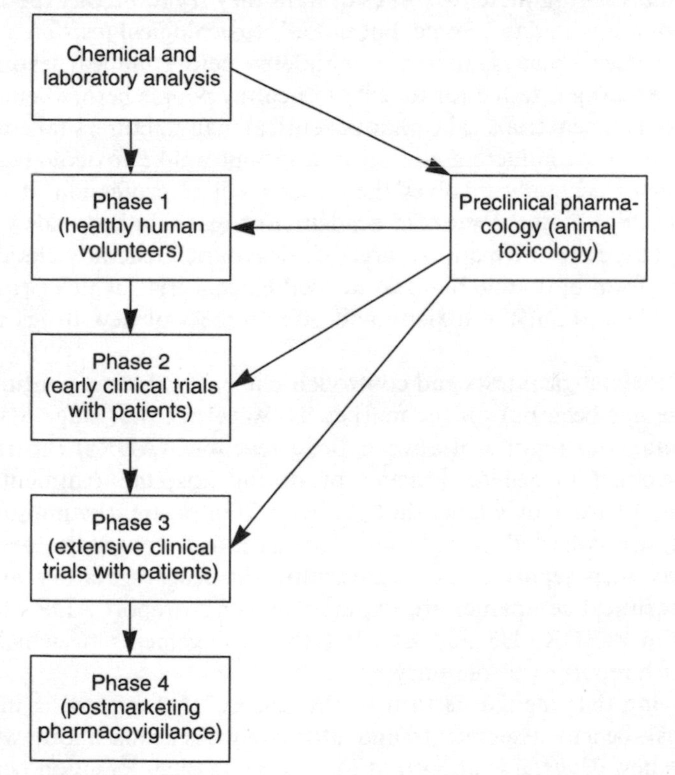

Figure 1 Drug development

scientific enterprises form an integral part of drug product evaluation procedures within the pharmaceutical industry and the governments that regulate it (Lumley and Walker 1985). But, increasingly, social scientists and consumer advocates are also drawing on these kinds of data in their analyses of medicines in society (Abraham 1995, Medawar 1992). In order to market new drugs, pharmaceutical companies have to comply with batteries of safety and efficacy tests set by the appropriate regulatory authorities. Both the safety and efficacy testing of drugs are characterised by considerable uncertainties. In this section I explain some of the fundamental uncertainties present in safety testing in order to illustrate the scientific context within which medicines regulation takes place

While a small amount of toxicology is undertaken in governments and universities, the toxicological testing of specific drug products is almost always conducted within the industry. In principle at least, the scientific objective of drug toxicology is to estimate the potential toxicity of pharmaceuticals to humans (Barnes and Denz 1954, WHO 1969, Zbinden 1987). Toxicologists draw on data derived from *in vitro* studies of cell behaviour carried out in glass dishes and from *in vivo* research conducted with whole live animals. Using these two types of data they try to predict the toxicity of a compound in humans. Some, but not all, toxicological tests on a new drug are pre-clinical, that is, they are completed before clinical testing begins. Also a new drug is tested for toxicity in healthy people before being given to patients in clinical trials. The pharmaceutical manufacturers take the lead in organising and conducting clinical testing, but unlike toxicological testing, the industry generally involves the wider medical profession at this stage, especially high status doctors in academia or teaching hospitals who might be sympathetic to the manufacturers' endeavours. Generally clinical testing takes the form of double-blind controlled clinical trials which provide comparative data about the toxicity and effectiveness of new drugs in patient groups.

Both toxicological tests and controlled clinical trials may continue after a medicine has been put on the market. However, at that stage, data, based on spontaneous reports of adverse drug reactions (ADRs) experienced by patients either in general practice or during hospital treatment, become available. Pharmacovigilance data, derived from postmarketing surveillance systems, are collected initially by pharmaceutical companies and doctors who may then report them to governmental drug regulatory authorities. Pharmaceutical companies are required by law to report ADRs to various degrees in the UK, US and EU, but the arrangements by which doctors make such reports are voluntary.

In saying that regulators turn to the science of drug testing in order to make risk-benefit assessments and ultimately decisions about whether to approve new drugs, it is important to appreciate the interaction between the science and politics of medicines regulation. To a large extent, pharmaceutical companies carry out such testing in order to meet the technical standards demanded of them by the regulatory authorities empowered to permit or withhold licence to marketing. In this sense, the regulatory science of drug testing has developed pragmatically to adjust to shifts in policy (Brown 1988, Lumley and Walker 1985, Schwartzmann 1976).

Moreover, a recurring feature of medicines regulation is that the underpinning science is characterised by considerable technical uncertainty. In toxicology, problems concerning the extrapolative validity of animal and cell studies to humans are substantial. In the case of testing whether a compound causes cell mutation, *in vitro* mutagenicity tests are employed, but these frequently yield inconsistent results and cannot, for example, determine whether a compound might induce cancer because only some cancers develop by genetic mechanisms (WHO 1974). For these reasons, it is supposed that *in vivo* lifetime toxicity tests with whole live animals are required to detect carcinogenic (*i.e.* cancer-inducing) compounds. However, on reviewing the pharmacology of six drugs in humans, dogs and rats, Litchfield concluded that 'many of the most serious side-effects that can result when a drug is given to man were not predictable from observations on dogs or rats' (1961: 34). More recently Salsburg (1983) provided a fairly comprehensive quantitative estimate of the validity of lifetime feeding studies of chemicals with rodents as tests for carcinogenicity and determined that they were more often wrong than right. Such fundamental uncertainties in toxicology have not been overcome in the 1990s (Personal Communication 1995).

Of course, with clinical trials there is no problem of extrapolation between species because drugs are administered to humans in such testing. However, only a small sample of the prospective population to be exposed to the drug can be studied in clinical trials. Furthermore, certain kinds of patients, who will probably be exposed to the drug if marketed, may be excluded from clinical trials because their multiple pathologies or other medication that they are taking might interfere with the scientific demands of controlled comparison. As Burley and Glynne put it: 'a great deal of artificial rigidity has necessarily to be built in [to clinical trials] which is at odds with normal clinical practice' (1985: 104). As a result, extrapolating findings about the safety and efficacy of medicines from clinical trials to the entire patient population is problematic. There may also be ethical reasons for such uncertainty. For example, it may be considered unethical to treat cancer or AIDS patients for very long on a new but toxic drug that has yet to prove its efficacy relative to other treatments available.

As regards the pharmacovigilance data collected after medicines are on the market, the difficulties are less about extrapolation and more about interpretative uncertainty. The fact that the reporting of ADRs by doctors is voluntary means that the proportion of real ADRs that such reports represent for any particular product is unknown, although in general terms it is thought that doctors report about five to ten per cent of such reactions (Walker and Lumley 1987). Nevertheless, such reporting is also thought to vary over time in ways that are not necessarily predictable. These problems mean that any attempt to make quantitative comparisons between the ADRs reported for different drugs is characterised by substantial uncertainty. To compound these difficulties the nature of the reports themselves merely indicates an association between the drug product and adverse reaction. The doctor may suspect that the drug caused the reaction but may not have established this. In the context of normal clinical practice patients may be suffering from many illnesses and be taking many drugs in addition to the suspected drug (Inman 1986). Under these conditions it is generally much more difficult to establish a causal link between a drug and adverse reaction than it is during clinical trials (Mann 1987). Hence, the extent to which pharmacovigilance provides valid and reliable data upon which regulatory authorities and others can reach assessments about medicines safety is limited.

To add further to the technical uncertainties underpinning medicines regulation, the three data sets, toxicology, clinical trials and pharmacovigilance may not produce consistent and confirmatory results, and indeed different scientific standards may be applied as to how consistencies between these very different types of data should be defined (Abraham and Sheppard 1995). These deep and extensive technical uncertainties in drug testing partly account for the fact that scientists in different national regulatory authorities can review the same data about the safety of a medicine and reach entirely contradictory regulatory decisions about it. This occurred most dramatically with *Halcion* (Abraham and Sheppard 1996, Gabe and Bury 1996), but has also happened with other drug products (Abraham 1995).

It is important to appreciate, however, that such uncertainties do not exist in a political vacuum. Pharmaceutical companies have real commercial interests in getting their drug products marketed quickly while patients have real health interests in receiving medicines that they need and that are maximally safe and effective. Sometimes these interests converge, but in other instances they may diverge or even conflict. In short, because the science base is so malleable, social and political factors may enter into medicines regulation *under the guise of technical problem-solving* with relative ease, and when they do so it is of crucial importance to understand how they relate to the competing interests involved. This is not to suggest that a more definitive regulatory science could expunge social and political factors, but it would make it less likely that they could be misrepresented as matters of technical calculation. Indeed, technical standards for drug testing are themselves set through negotiations between social and political interests. Implicit in such negotiations are social judgements about acceptable risk which are translated into bureaucratic demands on the science of drug testing. For example, there has been an extension of the safety testing of new drugs on healthy volunteers over the last decade. This is partly because of the problems of extrapolation from animal testing but also because of pressure from social interests to include more women and elderly people in safety testing (Abraham 1995, GAO 1992, Hamilton 1996). Furthermore, the nature of those social and political factors varies according to national, supranational and international context, with important sociological consequences for the interaction between regulatory institutions and medical science.

Medicines regulation in the UK

Prior to the mid-1960s, the British Government had trusted medical scientists in the pharmaceutical industry to develop and test drugs satisfactorily for safety and efficacy before putting them on the market without any regulatory review. However, after the thalidomide disaster came to light in 1961 the Government considered legislation to regulate drug safety with more urgency than ever before. In 1962 Enoch Powell, the Minister of Health, asked the Joint SubCommittee of the Standing Medical Advisory Committees under the chairmanship of Lord Cohen to advise on the testing and regulation of new drugs, but he also consulted the pharmaceutical industry. For, as the President of the Association of the British Pharmaceutical Industry (ABPI) pointed out to the Minister of State at a Board of Trade luncheon on 25 October 1962, the Ministry of Health was the industry's sponsoring department and had responsibility for the industry's scientific and commercial progress (PSGB 1962). For its part, the industry wanted some regulation to help rationalise its operations and to raise the standing of its products abroad, but it did not want a regulatory system too critical of its brand name products which were already established exports (Wheeler 1963, 1964). It was in this context that the Cohen SubCommittee sat down in consultation with the ABPI to advise the Ministry of Health on the kind of regulatory body which should set the tone for modern drug safety in the UK.

The Cohen Committee proposed that the Minister of Health should immediately and without legislation appoint an expert science advisory Committee on the Safety of Drugs (CSD), which would depend on the voluntary cooperation of the industry (Ministry of Health 1963). Accepting the advice of Cohen's Committee, the Health Ministers appointed the CSD with Sir Derrick Dunlop as chairman (PSGB 1963a). The degree of trust vested in the industry by these expert scientists is illustrated by Professor Wilson, a member of both the Cohen Committee and the CSD, who assumed:

> If a drug is shown to be harmful to animals, its use in Man is not contemplated, . . . and every reputable pharmaceutical firm and clinical investigator ensure to the best of current knowledge that all the appropriate

> investigations have been done before the drug is given to Man (Wilson 1962: 196).

The CSD was to invite reports on toxicity tests from the manufacturer, consider whether the drug should be put to clinical trials, obtain reports of such trials, and take into account the safety, efficacy and adverse effects of the drug (PSGB 1963b). It began operations on 1 January 1964 having pledged that information submitted to it by manufacturers about new drugs would be treated as confidential to the Committee ostensibly to ensure that the development of new drugs of therapeutic value was not hindered. Thus, before it had even begun regulatory activity the Committee sealed itself off from public scrutiny (PSGB 1963c). Moreover, regulatory review was deliberately rapid, averaging three months for new chemical entities and one month for novel reformulations (PSGB 1966). As the CSD commented in its annual report for 1966:

> it is fully recognised that a Committee such as this might exercise a detrimental effect on pharmaceutical research progress by unduly delaying the introduction of a possibly valuable drug or even by preventing its use altogether (PSGB 1967a: 59–60).

This approach, however, involved allowing the industry a very substantial amount of influence over the regulatory process. As Cahal, the CSD's Medical Assessor explained, the Committee was *dependent* on the industry's cooperation:

> One is often asked how the Committee manages to comply with its terms of reference with so small a staff. The answer is 'decentralisation', which means, since there is nowhere else to which we can decentralise, decentralisation to industry (US Congress 1970: 37).

Such extensive contact with industry, however, was not without impact on the way the CSD conducted itself, as a former member later explained:

> Looking back I see only one major error in our performance. We were so aware of the enormous cooperation that we received from the drug industry that the main Committee made every effort it could to see that submissions from firms were handled as rapidly as possible – as a result . . . the Adverse Reactions subcommittee and . . . the work of that subcommittee suffered (Wade 1983: 3).

In 1967 the Government outlined a product licensing system to regulate the quality, safety, and efficacy of new drugs that would replace the CSD (see Figure 2). The Minister of Health was to act as the licensing authority with a Medicines Commission to advise the Minister on the appointment of expert scientific committees (Cmnd 1967).

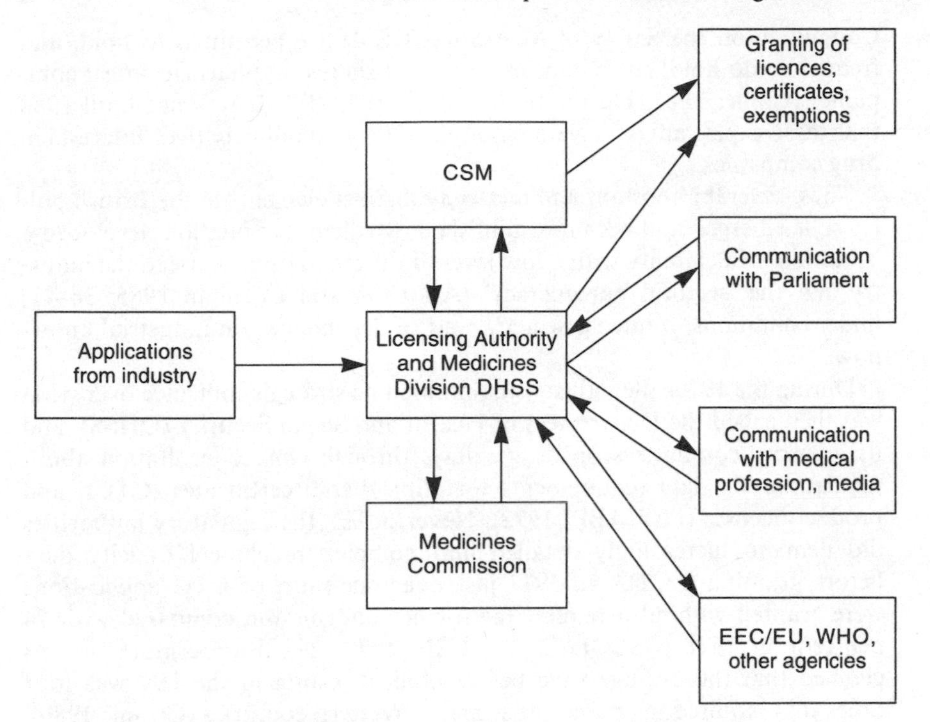

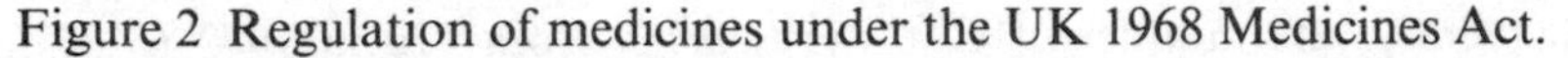
Figure 2 Regulation of medicines under the UK 1968 Medicines Act.

During the period leading up to this drug legislation the ABPI and the CSD maintained close liaison and their representatives responded with remarkable similarity to the issue of how the Government should regulate drugs. The ABPI supported the licensing system, provided that it was to be operated with the flexibility which characterised the CSD and did not impede the development of new medicines (ABPI 1968, PSGB 1967b). In March 1968 Dunlop asserted that laws enacted to assure drug safety and efficacy should not impose unnecessary restraints on the prosperity of the pharmaceutical industry, and noted that the medical profession depended on the industry's well-being (Dunlop 1971: 20, PSGB 1968a: 274–5).

On 15 February 1968 the Medicines Act, the basis for contemporary British medicines regulation, was passed in the House of Commons although it did not come into full operation until 1971 (PSGB 1968b). In May 1969, a few weeks after Dunlop had been appointed chairman of the new Medicines Commission, the ABPI were invited to have discussions with the Ministry of Health about the Commission's functions, structure and membership (ABPI 1970). Following the precedent set by the CSD, under the 1968 Medicines Act the entire decision-making process about specific drug product licence applications is secret and members of the Medicines Commission and its expert science advisory committee, such as the

Committee on the Safety of Medicines (CSM) are permitted to hold, and frequently do hold, consultancies with, and shares in, pharmaceutical companies (Collier 1985, Delamothe 1989, Scrip 1991a). It was not until 1988 that these expert advisers were required to declare publicly their interests in drug companies.

Thus, several important and relatively distinct elements in the British policy style emerged and became established. Medicines regulation developed a 'tradition' of flexibility with a low level of differentiation between the industry and the 'sectoral bureaucracy' (Atkinson and Coleman 1985: 36–41) under conditions of intense secrecy, and of dependence on industrial know-how.

During the 1970s the industry maintained its strategic influence over what was then called the Department of Health and Social Security (DHSS), and its advisory committees on drug safety, through close consultation about regulations on data requirements for clinical trial certificates (CTCs) and product licenses (Pls) (ABPI 1972). Nevertheless, the regulatory authorities did demand increasingly detailed and complex preclinical toxicity data before granting a CTC. In 1977 just over one-third of CTC applications were granted without a request for further information compared with 74 per cent in 1971 (CSD 1972: 12, CSN 1978: 28). Furthermore, it was claimed that the average time before clinical testing in the UK was four times that required in several other major Western countries (Cromie 1980). Consequently, the number of CTCs issued fell from 170 in 1972 to 87 in 1980 and, according to industrial representatives, companies shifted investment in clinical trials to locations outside the UK (ABPI 1977: 6, Griffin and Diggle 1981: 461).

On the other hand, stricter regulatory controls during this period were associated with fewer new chemical entities which reached the British market, but then ultimately failed to be of therapeutic value (Steward and Wibberley 1980: 119). This implied that the new regulations were performing at least some consumer protective function.

Throughout the late 1970s the ABPI complained about the extent of the DHSS's regulatory activity concerning preclinical data submissions for CTCs. Of particular significance was the conclusion reached in 1976 by the National Economic Development Council's 'sector working party' on the pharmaceutical industry. This was that for the pharmaceutical sector to maximise its contribution to a positive UK balance of payments through expansion of direct exports and import substitution, the DHSS should seek to have the minimum impact on the industry's research by ensuring that decisions regarding CTCs should be reached within two months (BMA 1976, PSGB 1976). Consequently, the CSM was asked to examine ways in which its procedures for assessing CTCs could be expedited (BMA 1976, PSGB 1976).

In 1979 a Conservative Government was elected with a monetarist philosophy and a commitment to reduce state intervention in the economy. This was later to be seen as the mark of a neo-liberal political agenda that would significantly affect the control and regulation of industry for at least the next two decades. By April 1980 Patrick Jenkin, then Secretary of State for Social Services, had announced the Government's intention to introduce a clinical trial exemption (CTX) scheme (ABPI 1980: 24). Thus, from March 1981 under this scheme an applicant needed only to submit a summary of the data relevant for a CTC to the regulatory authority, the Medicines Division of the DHSS, who then had five weeks to object to the proposed trials. If no objections were made in this time, a CTX was granted without reference to the CSM. But if an exemption were refused then, as before the CTX scheme, the applicant was required to submit all the relevant data for a CTC for review by both the Medicines Division and the CSM (ABPI 1980: 24, ABPI 1981: 6, DHSS MLX 130, DHSS MAl 62). The CTX scheme was much welcomed by the industry (Smart 1981). In August 1981 senior representatives of the Medicines Division stated clearly that the CTX scheme had become necessary 'because early developmental work on new drugs was going abroad to the detriment of British industry and with a loss of skill in our departments of pharmacology', thus indicating that the regulatory authority had adopted both the industry's interpretation of the effects of the CTC requirements and its suggestions for change (Griffin and Long 1981: 477).

The industry continued to complain about the delays in the licensing process throughout the early and mid-1980s. This prompted a Government review of medicines control conducted by two managerial advisers, John Evans and Peter Cunliffe. In 1988 they proposed several organisational changes aimed at increasing the 'efficiency' of British medicines regulation. In particular they took up the pharmaceutical industry's suggestion that it would be willing to pay the cost of funding medicines approval if that were to result in a more 'efficient service' (Scrip 1988b: 3). Previously the regulatory authority had been funded 65 per cent by licensing application fees from the pharmaceutical industry and 35 per cent by the Government via taxes (Scrip 1988a: 24). Following the Evans/Cunliffe Report in 1989, and exemplifying a more general trend under the Conservative Government, the Medicines Division became the Medicines Control Agency (MCA), almost entirely funded by industry fees. It is now essentially run as a business selling its regulatory services to the industry and promoting itself as the fastest licensing authority in the world for new chemical entities (Scrip 1991b: 2).

It is interesting to note that until mid-1988 product licences were suspended or revoked by the regulatory authorities at an average rate of about ten per year. Yet over the next three years no drug lost its licence until Upjohn refused to withdraw *Halcion* (Medawar 1992: 245). Certainly the switch to the business-like MCA does not seem to have improved public accessibility to the regulatory process, for there is now a £250 admission fee to its annual meetings which are ostensibly held to 'hear and take account of the views of those for whom it provides services' (MCA 1991). As that fee almost ensures that only company representatives will be able to attend, this implies that the regulatory authority regards the industry, rather than consumers as its primary constituency of service. Such developments coincided with the Department of Health's failure to support the Medicines Information Bill put forward by a Member of Parliament in 1993. The Bill proposed to make British medicines regulation more transparent and to provide consumers with rights of access to information about the bases for regulatory decisions, but it failed to get through Parliament.

The foregoing analysis suggests that the pharmaceutical industry has been successful in obtaining from the British Government the kind of regulatory authority most suited to their interests. This is not to suggest that at all times, and in every case, the industry has been able to extract whatever agreements and concessions it has demanded. During the mid-1970s, British regulators exhibited behaviour less convergent with industrial interests, perhaps encouraged by a Labour Government, which was initially rather critical of the pharmaceutical industry. Nevertheless, to a good approximation, British medicines regulation is characterised by industrial 'corporate bias' (Middlemas 1979, Lexchin 1990). This involves regular bargaining between organised interests (the pharmaceutical industry) and the State about the extent of regulation, and importantly goes beyond 'pressure pluralism' (Atkinson and Coleman 1985: 41), such as lobbying and pressure group politics:

> To put it simply, what had been merely interest groups crossed the political threshold and became part of the extended State; a position from which other groups, even if they too held political power, were excluded (Middlemas 1979: 373).

Medicines regulation in the US

Since the enactment of the 1938 Food, Drug and Cosmetic Act, pharmaceutical manufacturers in the US have been required to obtain permission from the American drug regulatory authority, the Food and Drug Administration (FDA), to market a new drug. Following Congressional hearings conducted by Senator Kefauver in the late 1950s which exposed the pharmaceutical industry to embarrassing criticism, medicines regulation was substantially tightened up with the 1962 Kefauver-Harris Amendments to the 1938 Act (Dameshek 1960, Harris 1964: 96–104, 220). The most significant aspects of the 1962 Drug Amendments concerned requirements for efficacy rather than safety. For example, they enabled the FDA to require manufacturers to provide 'substantial evidence of effectiveness' consisting of 'adequate and well-controlled investigations' (Harris 1964: 204–5). The FDA also acquired the authority to withdraw approval already granted to drugs if they were considered unsafe or lacked 'substantial evidence of effectiveness' (Temin 1980: 125, Silverman and Lee 1974: 121).

During the 1960s the FDA was sluggish about its new powers to remove ineffective drugs from the market. However, in cases such as the controversy over the withdrawal of the combination drug, *Panalba*, when industrial interests seem to have unduly influenced the decision-making process of the regulatory authorities, oversight procedures by Congress have exposed such industrial bias and reduced its impact (US Congress 1969: 199–201). Under the 1938 Act the FDA is specifically charged by Congress with the responsibility to protect the public from unsafe drugs. Moreover, in 1967 the US Freedom of Information Act was passed. This Act provides members of the public with rights of extensive access to the documentation recording the FDA's decision making about approving new drugs, although that access only comes into effect after the drug has been approved. This includes access not only to the internal medical assessments by FDA scientists but also to the records of the FDA's scientific advisory committee meetings and meetings with drug manufacturers. Consequently the drug regulatory authorities in the US operate in a political environment in which there is much more explicit recognition of potentially conflicting interests between manufacturers and consumers and more opportunity for consumers to scrutinise how well the regulators are protecting their interests than exists in the UK. These factors, combined with the fact that the US is a much more litigious society than the UK, have usually led to a more adversarial relationship between the regulatory authorities, the pharmaceutical industry and consumer organisations.

This is not to suggest that the FDA is immune from industrial capture. For example, in the early 1970s under the Nixon Administration the 'industry-friendly' Charles Edwards was appointed as Commissioner of the FDA. During this period FDA management instigated its policy of 'neutralising' medical scientists within the organisation who were 'adversarial' towards industry. This was revealed in 1974 when a Democrat-led Congressional Committee heard remarkable testimonies from nine reviewing officers at the FDA's Bureau of Drugs and two physicians who had previously worked in the Agency. All were, or had been responsible for reviewing some part of the data submitted by drug companies in order to get their products approved for marketing. These medical scientists claimed that when they recommended approval of a drug their analyses were rarely challenged but their recommendations for non-approval were unjustifiably over-ruled. Many of them testified that when they insisted on recommending non-approval they experienced harassment within the Agency and were sometimes removed from reviewing the particular drug in question (US Congress 1974: 2830).

Significantly, Congress set up a Special Panel to investigate allegations of 'undue industry influence', 'improper transfers, details or removals', 'improper use of advisory committees' and 'improper use of medical officer recommendations' (Review Panel on New Drug Regulation 1976: 16). In 1977 the Panel concluded that, although the FDA had not been 'dominated' by the pharmaceutical industry during the period, 'inappropriate contacts with drug companies occurred'. Furthermore, according to the Panel, from 1970 the FDA's management established and sought to implement a deliberate policy of 'making the Agency less adversarial towards and more co-operative with drug manufacturers, and to neutralise reviewing medical officers who followed a different philosophy' such as those who gave testimony to Congress (Review Panel of New Drug Regulation 1977: 2). As to the method of 'neutralisation', the Panel revealed:

> The program to neutralise the more adversarial reviewers was carried out by various devices, including a systematic pattern of involuntary transfers to positions which the incumbents did not want, and in a few cases removal from the review of particular drugs. FDA management generally concealed the truth about the reasons for the transfers from the persons affected (Review Panel of New Drug Regulation 1977: 2).

In an attempt to counteract such industrial capture of the FDA, Congress introduced the 1978 Ethics in Government Act, which prohibits FDA scientists from moving within two years to an industry they regulated while at the FDA. This Act also served to reinforce the existing rules governing appropriate contacts by FDA regulators with industry. Congress, therefore, sought to re-establish a more adversarial relationship between the FDA and pharmaceutical manufacturers than was evident in the early 1970s. Moreover, it was assisted in this task by the Carter Administration's appointment of Donald Kennedy as FDA Commissioner, a staunch critic of the pharmaceutical industry.

During the Reagan and Bush Administrations of the 1980s, however, the FDA was continually under pressure from the Executive to limit its regulatory activities and to avoid hampering industrial competitiveness. In particular, Kennedy was removed and the FDA was asked to accelerate its approval rates, often with fewer resources than before. This reflected the strength of the neo-liberal political agenda during the 1980s and into the 1990s in the US as well as in the UK and other industrialised countries. As a consequence, it became of increasing political importance for regulators to show that they could approve new drugs faster than their counterparts in other industrialised nations, and to search for ways of so doing.

In this context, the FDA made a number of regulatory decisions which were highly controversial and considered to be unsatisfactory by a Democratic Congress and public interest groups. During the 1980s, Democrat-led Congressional Committees investigated in depth the FDA's approval of specific drug products on no less than five occasions (US Congress 1982, 1983, 1986, 1987, 1988). All of these drugs, except for one, were also approved in the UK, but British regulators were never required to defend those approvals before a Parliamentary Select Committee. By contrast, in each case FDA medical scientists were subjected to intense scrutiny that required them to justify their risk-benefit assessments under close questioning in public. Such investigations reminded FDA regulators that they might be called to account in public for their decisions and to demonstrate that these were in consumers' interests.

The attempts at industrial capture of US medicines regulation, aided in no small measure by the neo-liberal politics mentioned above, have been only partly successful because of the extent to which the regulatory authorities can be made publicly accountable either by Congress or consumer interest groups using the Freedom of Information Act to build a case in the Courts. Of particular significance in the US is the legislative-executive competition for control over the bureaucracy (FDA) between Congress and the Administration. This tends to amplify legislative and procedural controls on the bureaucracy compared with the situation in the UK where the State's executive and legislative functions are fused into a cooperative arrangement within the governing majority Parliamentary Party. This means that FDA medical scientists tend to be more cautious about embracing industrial perspectives and less permissive about drug products than their British counterparts, although one might expect this tendency to be minimised when there is a neo-liberal Administration *and* Congress. Moreover, the FDA attracts medical scientists who are likely to hold more critical views about the pharmaceutical industry than scientists serving the British regulatory authorities, because of its greater institutional independence from the industry.

Europeanisation of medicines regulation

Considerable national differences in pharmaceutical markets exist between Member States of the European Union (EU), formerly the European Community (EC). For example, of the first 50 pharmaceutical products most sold by value in 1992 in France, Germany, Italy and the UK, only seven were common to all four lists (Garattini and Garattini 1993). This is partly a consequence of different regulatory decisions about approving drug products in the four countries. The Europeanisation of medicines regulation is the process resulting from efforts by industry and national Governments within the EC/EU to harmonise standards of regulatory evaluation in order to accommodate the prospect of a single European market in pharmaceuticals. By harmonising such standards the intention is to reduce and eventually eliminate conflicting regulatory decisions by different national regulatory authorities in the EU. This process dates back to 1965 when the EC first put forward a Council Directive that made provision for the regulation of medicinal products in the Community (Council of Ministers 65/65/EC), but it has acquired greater urgency during the 1990s because of a growing concern among European governments and industrialists that an integrated EU-wide pharmaceutical market is a necessary condition for the effective competitiveness of European firms on the world stage (Hancher 1996: 179–80).

This process of Europeanisation has inevitably involved the moulding of medical science in the field of drug testing. In 1975, following a period of immense disagreement between national regulatory authorities, the European Commission issued Directives that stipulated common technical standards to be applied across the Community and established its own committee of European expert scientists, known as the Committee for Proprietary Medicinal Products (CPMP) (Hancher 1990), Cartwright and Matthews 1991, Council of Ministers 75/319/EC). These Directives included requirements concerning the conduct of clinical trials, such as minimum numbers and age distribution of patients, details of patient withdrawals from the trial, and frequency and statistical analyses of patients' adverse effects (Council of Ministers 75/318/EC). By 1983 the EC was producing Council Recommendations that provided preferred standards for conducting carcinogenicity tests of drugs on animals. These included guidance on the number and strains of animal species, the relationship between dosage to animals and intended therapeutic dosage in humans, the age of the animals, the number of treatment groups, the number of animals in the control and treatment groups, the duration of the studies, the desired survival rate in the control group, and the satisfactory conduct of autopsies by identifying, enumerating and grading various types of tumours wherever possible (Council of Ministers Recommendation 83/571/EC).

In addition, to further facilitate Europeanisation of medicines regulation, in 1985 the Multi-State procedure was established. This enabled manufacturers to seek simultaneous authorisation to market a product in two or more Member States providing that that manufacturer had an existing authorisation in at least one Member State. Under this procedure once a drug product was approved in one Member State the manufacturer could ask the regulators in that Member State to act as rapporteur, and to send the application and approval details to two or more other Member States. These recipient Member States were then expected to give due respect to the approval decision by the rapporteur, but if they disagreed then the CPMP was called upon to adjudicate in the spirit of harmonisation. However, the CPMP's decision was non-binding and so recipient Member States could continue to dissent and did so (Medicines Control Agency 1993: 47).

In fact, considerable disagreement plagued the Multi-State procedure with the result that harmonisation was very limited. It did not, therefore, serve the industry's interests in gaining more rapid and efficient access to a European-wide market. Consequently, the procedure was used by pharmaceutical companies for just one per cent of the total applications in Germany and just three per cent in the UK (Jeffereys 1993: 33). However, it should be noted that this situation may be changing because since 1995, under the 'decentralised procedure', which operates on the same principles as the Multi-State procedure and has succeeded it, the decisions of the CPMP have become *binding* on Member States. The other major change under the decentralised procedure is that the CPMP is now supported by a scientific secretariat, the European Medicines Evaluation Agency (EMEA) in London. Thus, despite efforts to standardise the underpinning of medicines regulation in the EU, disputes continued because the interpretation of the significance of those standards for regulatory decisions varied according to social and political judgements about acceptable risk/benefit ratios. Furthermore, such Europeanisation has not eliminated the scientific uncertainty of drug testing but merely standardised it.

Initially, the pharmaceutical industry was worried that the most stringent factors adopted by each Member State might be aggregated by the CPMP. The Committee, however, responded by claiming that it had adopted attitudes that moderated rather than aggregated the demands of all its Members (Abraham and Charlton 1994: 6–7). It may be concluded from this that the safety standards adopted by the CPMP have been more relaxed than those of some Member states in certain respects. In short, the goal of mutual recognition via the Multi-State procedure and its successor, put some national regulatory authorities under pressure to accept what they regard as suspect evaluations by other Member States. The regulators, such as those at the Swedish Medical Products Agency, who believe that they apply rigorous safety standards on a national level are concerned that Europeanisation is compromising those standards (Abraham and Charlton 1994).

The EMEA is sometimes likened to a European FDA, but it is much more accurate to liken the CPMP to a European CSM. This is because the Europeanisation of medicines regulation is being modelled much more on the British system than the US system. Scientific and regulatory standards are the products of continual corporatist negotiations between industry and the regulators with limited input from elsewhere. For example, the rapporteur arrangements generally favour industry. Manufacturers choose rapporteurs with whom they have 'good relations' and 'good connections' to assist in gaining wider European approval for their products. During interviews some industrialists openly acknowledged that by choosing a sympathetic rapporteur manufacturers hope to slant the regulatory scientific assessment in their favour (Abraham and Charlton 1994).

In contrast to such privileged access to regulators by industry, consumer and 'public interest' groups are only minimally consulted. There are no rights to access to information concerning the internal decision making of the EMEA or the CPMP. Such data are regarded as confidential. Since 1996, for biotechnology drug products and innovative new chemical entities, the EMEA has made publicly available a European Product Assessment

Report (EPAR). This is a document of about 30 pages which provides licensing information similar to the American Summary Basis of Approval (SBA). However, even this modicum of information remains unavailable for the vast majority of drug products reviewed by the EU (Abraham and Lewis 1997). In the meantime the EU systems of medicines regulations are only marginally more transparent than the British system. Swedish consumers, who are accustomed to freedom of information rights, are likely to feel the costs of such secrecy more than the citizens of other EU countries (Carvel 1994), but at the time of writing there is little sign that the Swedish MPA has the political will to prevent a culture of secrecy becoming a hallmark of the Europeanisation of medicines regulation.

Globalisation of medicines regulation

During the late 1980s the CPMP circulated proposals for a flexible pre-clinical medicines testing regulatory strategy which implied that particular tests would be required only if there was a high probability that the test compound would be toxic in that respect. The industry welcomed the idea because it would reduce the costs of drug development, but did not think it was workable until there was greater global harmonisation of toxicological testing requirements (Abraham and Charlton 1994). Thus, the industry's drive for regulatory harmonisation in order to maximise access to markets with the most rationalised drug testing demands does not stop at Europeanisation.

This partly explains the emergence of the International Conference on Harmonisation (ICH) which was first convened in November 1991. The ICH is 'sponsored' by the regulatory authorities and pharmaceutical trade associations of the US, Japan and the EC/EU. It has established itself as a major force in setting the agenda for toxicological testing requirements (Cone *et al.* 1992). While regulators attend ICH meetings, it is an industry-led organisation because it is industrialists who are driving the global harmonisation process, in an attempt to increase industrial efficiency by reducing the burden of toxicological testing. Nevertheless, the enthusiastic participation in, and frequent support for, ICH developments among regulators reveals the extent to which the neo-liberal political agenda has formulated and consolidated a corporatist, rather than adversarial, model of regulation transnationally.

This has a marked impact on the science of drug testing. For example, the ICH has been influential in persuading regulators in Japan and the US to fall into line with the EC/EU by abandoning 12-month studies in rodents as a routine toxicological requirement because, according to EU industrialists and regulators, such studies provide little additional safety information. However, corporatism is not hegemonic. The ICH was not able to persuade the FDA to discard its requirement for 12-month studies on dogs for medicines intended for chronic use (Parkinson 1992). This was manifest during a controversy between the UK-based Centre for Medicines Research (CMR), funded by the ABPI, and the FDA. The FDA disputed the CMR's assertion that the 12-month dog studies did not add any significant safety data.

The controversy illustrates how the institutional interests of industry and regulators expressed in harmonisation go right to the heart of the scientific uncertainty inherent in drug testing. Lack of consensus stemmed from the fact that the FDA defined significant findings as 'clinical significant toxicity . . . serious enough to influence the design or progress of clinical trials' and necessitating 'regulatory action based on a reduced margin of safety' (Contrera *et al.* 1993: 65). By contrast, the CMR defined findings as 'significant' if they would 'influence the development of a compound' (Lumley *et al.* 1993: 54). Hence, the FDA defined 'significant' in terms of safety concerns and the need to intervene on the basis of those concerns, thus revealing the Agency's awareness of its very public political environment of Congressional oversight, litigation and active consumer organisations. On the other hand, the CMR defined it in terms of the closed commercial world, that is, whether the findings were so damaging to a prospective product that its development was put in jeopardy. Evidently these different institutional perspectives influenced the definition of satisfactory standards for the science of drug testing.

Another important element in the debate about the reduction of animal toxicology testing is the role of the animal protection movement. The mass media frequently present this social movement and the pharmaceutical industry as adversaries. However, since 1986 the industry has given 'tacit support' for the moderate elements of the movement, such as the Fund for the Replacement of Animals in Medical Experiments (FRAME), which, in 1996 secured £1 million from the pharmaceutical and cosmetic industries to research alternatives to animal testing (Elston 1996: 14, 16–17). Thus, there is evidence of the pharmaceutical industry's willingness to be associated with some of the arguments of the animal protection movement that imply a need to reduce animal toxicology testing. For example, Glaxo sponsored the FRAME publication, *Developing Alternatives to Animal Experimentation*, which lists economic saving alongside welfare of animals as advantages of alternatives to animal toxicological tests. Clearly, the alteration of regulations is an important goal. The authors of that publication, Fentem and Balls comment:

> There are prospects that some of the animal toxicity tests currently required by law could be partly or fully replaced in the foreseeable future (1993: 6).

Furthermore, in the early 1990s, one of these co-authors, who 'acted as principal adviser to the British Government during the passage of the

Animals (Scientific Procedures) Act 1986', was Secretary to the (FRAME) Toxicity Committee, whose membership has also included the Director of the CMR (FRAME 1991: 116, 118). This is part of what Elston (1996) describes as a growing 'middle of the road' alliance between industry and the moderate wing of the animal protection movement, involving the 'incorporation' of the former by the latter (Gottweis 1995). Conversely, to a considerable extent, animal protection interests have been 'incorporated' into industrial strategies towards regulatory toxicology, including ICH. Whether this is in the best interests of consumers and patients remains to be seen.

The ICH has shown little sign of developing harmonisation resulting directly in raising safety standards. Also underdeveloped within ICH work are discussions about comparative efficacy, which might lead to more efficient drug development, but also probably higher standards of efficacy testing for the industry and, therefore, some cost increases. Thus, the efficiency drive of the ICH impacts upon the science of drug testing with a particular flavour and with considerable success by its own criteria. However, such success may be because of its exclusion of wider social groups in discussions of the development of drug testing. Issues and priorities that might provoke dissension in the wider social environment are transformed into areas of agreement by the bilateral industry-regulator context of the ICH. The disagreements between the FDA and the CMR are indications of just how much scope there could be for divergent opinions about scientific standards if an even broader constituency of perspectives on drug safety were accommodated within these debates about regulatory medical science.

Conclusions and implications

Throughout the industrialised world the regulation of medicines draws on the sciences/technologies of toxicology, clinical trials and epidemiology. Thus, scientific authority may be seen as alive and well in this field. Gabe and Bury (1996) argue that a major change occurring in contemporary societies is that medical authority is 'fracturing', and that the differences between the lay and the expert are becoming blurred. While this is probably true in cases of medical controversy that spill over into the public domain (*e.g. Halcion*), such cases need to be set alongside the increasing trend towards harmonisation and standardisation of medical science aimed at increasing expert consensus. So long as those scientific developments can be kept away from the awkward scrutiny of public interest groups, drug product liability lawyers and the mass media, then convergence, rather than fragmentation, of medical expertise is likely to be the dominant trend. This is reflected in the technocratic style of regulatory science adopted by the EU. The confidential rapporteur system adjudicated by expert medical scientists on the CPMP is, in part, an attempt to regulate by scientistic peer review.

Perhaps of more significance is the fact that the emergence of Europeanised medicines regulation has taken the form of 'corporatist' political activity involving regular bargaining between organised industrial interests and the emerging European State (Cawson 1986). It is no coincidence that consensus about medicines regulation is much less marked in the US than in the UK or the EU. It is also clear that the FDA is much more cautious about harmonising drug testing standards downwards than its counterparts in Europe. I have argued that this has less to do with the underpinning of science, and much more to do with the political environment of drug regulation in the US, which is less corporatist and more unpredictable for industry. This said, it is important to appreciate that micro-sociological processes can produce anomalies within these general trends. To mention a couple of examples: during the late 1980s it was consumers/patients more than industry who wanted anti-AIDS drugs to be approved quickly (Walker 1993); and in the 1990s the British regulatory authorities appear to have applied more rigorous safety standards to *Halcion* than the FDA (Abraham and Sheppard 1996).

It follows from the above analysis that the best consumer protective policies are likely to be those that maximise the transparency of regulatory decisions, perhaps by freedom of information laws or by active critical legislative oversight or both. Furthermore, greater transparency might benefit the development of toxicological science whose findings are currently being under-utilised. This is recognised by industrialists as well as consumer organisations. For example, during interviews, two senior industrial toxicologists responded to the failure of the British Medicines Information Bill in 1993 as follows:

> Scientist A: The industry, as a whole, thought it was a very good idea that the Bill got kicked out. But I mean for a toxicologist it would be very useful if we could look up other people's data.
>
> Scientist B: Absolutely, because we could learn from that and I think that, overall medicines would be a lot safer if we could access other people's data (Abraham and Charlton 1994: 13).

Evidently, industrial scientists do not necessarily share the perspectives of their employing institutions regarding the development of toxicology. Where toxicologists are more independent from industry, such as the academics conducting basic research for the British Medical Research Council, there may be even greater divergence of opinion away from the industrial instrumentalism that permeates regulatory toxicology.

There is also a need to widen public participation in the work of organisations such as the ICH so that the agenda for developments in the science of drug testing can be broadened to examine the comparative need for prospective drugs. In particular, the European harmonisation efforts put too much emphasis on industrial efficiency and not enough on accountability to non-industrial interests. The EU should examine ways of encouraging such participation. It should be noted, however, that this would not always result in greater support for the maximisation of patients' interests. For example, some animal protection interest groups might oppose greater consumer protection if it implied regulatory demands involving more extensive animal toxicological testing.

This research confirms the importance of work in sociology that takes account of the role of social, political and economic interests in influencing the direction of medical knowledge. It also suggests that sociology of health and illness has much to benefit from taking account of drug development and regulation. Trends in iatrogenic disease cannot be fully understood without reference to the regulatory science in the pharmaceuticals sector. The information about medicines which general practitioners may pass on to their patients is fundamentally defined and circumscribed by medical scientists working in government and industry, whose efforts are, in turn, framed by the interests of state, economy and other significant social forces. That information may be a central feature of the meaning that the patient attributes to the medication, with consequences for adherence and use. Regulatory science affects patients in clinical trials very directly. In this case patients' experiences of trial drugs are themselves part of the construction of regulatory science. Hence, there is a fusion of medical sociology with the sociology of regulatory science. Finally, the prospects of patients gaining more knowledge about, and control over, the management of their own treatment will be partly determined by their interest organisation and representation within the regulatory process. As has been shown in the case of AIDS, patient organisation can have an impact on regulatory science and the direction of drug development (Epstein 1995, Walker 1993). In other words, medical sociology can open up many fruitful routes of inquiry by appreciating that the medicines which patients use, how they use them, and the meanings they attribute to them influence, and are influenced by, processes that are divorced in space and time from doctor-patient relations and the actual experiences of use.

Acknowledgements

I am grateful to Mary Ann Elston and two anonymous referees for their comments on previous versions of this paper. I am also grateful to the ESRC for funding research which forms the basis for part of this chapter.

References

Abraham, J. (1995) *Science, Politics and the Pharmaceutical Industry: Controversy and Bias in Drug Regulation*. London/New York: UCL/St Martins Press.

Abraham, J. and Millstone, E. (1989) Food additive controls: some international comparisons, *Food Policy*, 14, 43–57.

Abraham, J. and Charlton, N. (1994) *The Europeanisation of Regulatory Toxicology*. Swindon: ESRC Report.

Abraham, J. and Sheppard, J. (1995) *Expert and Public Assessments of Medicines Safety*. Swindon: ESRC Report.

Abraham, J. and Sheppard, J. (1996) *Conflicting Scientific Expertise in British and American Medicines Control*. Swindon: ESRC Report.

Abraham, J. and Lewis, G. (1997) *The Interaction between European Medicines Regulation and Toxicological Science*. Swindon: ESRC Report.

Association of the British Pharmaceutical Industry (ABPI) (1968) Legislation: Medicines Bill, *ABPI Annual Report 1967–8*. London: ABPI.

Association of the British Pharmaceutical Industry (ABPI) (1970) Legislation: Medicines Commission, *ABPI Annual Report 1969–70*. London: ABPI.

ABPI (1972) Review of the Year: Medicines Act 1968, *ABPI Annual Report 1971–72*. London: ABPI.

ABPI (1977) Lessons of a decade, *ABPI News*, 164, 6.

ABPI (1980) Annual dinner 1980, *Annual Report 1979–80*, 24.

ABPI (1981) Medical and scientific affairs, *ABPI Annual Report 1980–81*, 6.

Atkinson, M.M. and Coleman, W.D. (1985) Corporatism and industrial policy. In Cawson, A. (ed) *Organized Interests and the State*. London: Sage.

Barnes, B. and Shapin, S. (1979) Introduction. In Barnes, B. and Shapin, S. (eds) *Natural Order: Historical Studies of Scientific Culture*. London: Sage.

Barnes, J.M. and Denz, F.A. (1954) Experimental methods used in determining chronic toxicity: a critical review, *Pharmacological Reviews*, 6, 191–242.

Bartley, M. (1990) Do we need a strong programme in medical sociology? *Sociology of Health and Illness*, 12, 371–90.

Bloor, D. (1973) Wittgenstein and Mannheim on the sociology of mathematics, *Studies in the History and Philosophy of Science*, 4, 173–91.

Bhaskar, R. (1979) *The Possibility of Naturalism*. Brighton: Harvester Press.

Bhaskar, R. (1986) *Scientific Realism and Human Emancipation*. London: Verso.

Brickman, R., Jasanoff, S. and Ilgen, T. (1985) *Controlling Chemicals: the Politics of Regulation in Europe and the United States*. Ithaca, NY: Cornell University Press.

British Medical Association (BMA) (1976) Pharmaceutical industry, *British Medical Journal*, 4 December, 1397.

Brown, V.K. (1988) *Acute and Sub-Acute Toxicology*. London: Edward Arnold.

Bunge, M. (1991) A critical examination of the new sociology of science, *Philosophy of Social Sciences*, 21, 524–60.

Bunkle, P. (1993) Calling the shots? The international politics of depo-provera. In Harding, S. (ed) *The Racial Economy of Science*. Bloomington: Indiana University Press.

Burley, D.M. and Glynne, A. (1985) Clinical trials. In Burley, D.M. and Binns, T.B. (eds) *Pharmaceutical Medicine*. London: Edward Arnold.

Cartwright, A.C. and Matthews, B.R. (eds) (1991) *Pharmaceutical Product Licensing: Requirements for Europe*. London: Elliss Harwood.

Carvel, J. (1994) Sweden plans law to blunt EU secrecy, *The Guardian*, 9 November.

Cawson, A. (ed) (1985) *Organised Interests and the State: Studies in Meso-Corporatism*. London: Sage.

Cawson, A. (1986) *Corporatism and Political Theory*. Oxford: Basil Blackwell.

Chubin, D. and Restivo, S. (1983) The 'mooting' of science studies: research programmes and science policy. In Knorr-Cetina, K.D. and Mulkay, M. (eds) *Science Observed*. London: Sage.

Cmnd 3395 (1967) *Forthcoming Legislation on the Safety, Quality and Description of Drugs and Medicines*. London: HMSO.

Collier, J. (1985) Licencing and provision of medicines in the UK: an appraisal, *Lancet*, 17 August, 377–80.

Committee on Safety of Drugs (CSD) (1972) *Report for Year Ending 1971*. London: HMSO.

Committee on Safety of Medicines (CSM) (1978) *Annual Report 1977*. London: HMSO.

Cone, M., Couper, M.R., Dunne, J.R. and Thomas, M. (1992) Harmonisation of international registration requirements for pharmaceuticals. In Griffin, J.P. (ed) *Medicines: Regulation, Research and Risk*. Belfast: Queen's University Press.

Contrera, J.F., Degeorge, J. and Jacobs, A.C. (1993) A retrospective comparison of the results of 6 and 12 month non-rodent toxicity studies, *Adverse Drug Reaction Toxicology Review*, 12, 63–76.

Council of Ministers Directive 65/65/EC *On the approximation of provisions laid down by law, regulation or administrative action relating to proprietary medicinal products*.

Council of Ministers Directive 75/318/EC *On the approximation of the laws of Member States relating to analytical, pharmaco-toxicological and clinical standards and protocols in respect of testing of proprietary medicinal products*.

Council of Ministers Directive 75/319/EC *On the approximation of provision laid down by law, regulation or administrative action relating to proprietary medicinal products*.

Council of Ministers Recommendation 83/571/EC *Recommendation concerning tests relating to the placement on the market of proprietary medicinal products*.

Cromie, B.J. (1980) Testing new drugs in the UK, *Journal of the Royal Society of Medicine*, 73, 379–80.

Dameshek, W. (1960) Chloramphenicol: a new warning, *Journal of the American Medical Association*, 174, 1853.

Davis, P. (1996) (ed) *Contested Ground*. Oxford: Oxford University Press.

Delamothe, T. (1989) Drug watchdogs and the drug industry, *British Medical Journal*, 299, 476.

Department of Health and Social Security (DHSS) *MLX*, 130.

Department of Health and Social Security (DHSS) *MAL*, 62.

Dunlop, D. (1971) *The Problem of Modern Medicines and Their Control*. Twelfth Maurice Bloch lecture, 11 February. Glasgow: University of Glasgow.

Elston, M.A. (1994) The anti-vivisectionist movement and the science of medicine. In Gabe, J., Kelleher, D. and Williams, G. (eds) *Challenging Medicine*. London: Routledge.

Elston, M.A. (1996) *The Public Controversy over Animal Experimentation since 1960: End-of-Grant Report*. Swindon: ESRC.

Epstein, S. (1995) The construction of lay expertise: AIDS activism and the forging of credibility in the reform of clinical trials, *Science, Technology and Human Values*, 20, 408–37.

Fentem, J. and Balls, M. (1993) *Developing Alternatives to Animal Experimentation*, Glaxo Group Research. Cambridge: Hobsons.

FRAME (1991) Animals and alternatives in toxicology: present status and future prospects: The Second Report of the FRAME Toxicity Committee, *Alternatives to Laboratory Animals*, 19, 116–38.

Gabe, J. and Bury, M. (1996) Halcion nights: a sociological account of a medical controversy, *Sociology*, 30, 447–69.

Garattini, S. and Garattini, L. (1993) Pharmaceutical prescriptions in four European countries, *Lancet*, 342, 1191.

General Accounting Office (GAO) (1992) *Women's Health: FDA Needs to Ensure more Study of Gender Differences in Prescription Drug Testing*. GAO/HRD-93-17. Washington, DC: GPO.

Gillespie, B., Eva, P. and Johnston, R. (1979) Carcinogenic risk assessment in the United States and Britain; the case of aldrin/dieldrin, *Social Studies of Science*, 9, 265–301.

Gottweis, H. (1995) German politics of genetic engineering and its deconstruction, *Social Studies of Science*, 25, 195–235.

Greenwood, J. and Ronit, K. (1991) Pharmaceutical regulation in Denmark and the UK: reformulating interest representation to the transnational level, *European Journal of Political Research*, 19, 327–59.

Griffin, J.P. and Diggle, G.E. (1981) A survey of products licenced in the UK from 1971–81, *British Journal of Clinical Pharmacology*, 123, 453–63.

Griffin, J.P. and Long, J.R. (1981) New procedures affecting the conduct of clinical trials in the United Kingdom, *British Medical Journal*, 283, 477.

Hamilton, J.A. (1996) Women and health policy: on the inclusion of females in clinical trials. In Sargent, C.F. and Brettell, C.B. (eds) *Gender and Health: an International Perspective*. New Jersey: Prentice-Hall.

Hancher, L. (1990) *Regulating for Competition*. Oxford: Clarendon.

Hancher, L. (1996) Pharmaceutical policy and regulation: setting the pace in the European Community. In Davis, P. (ed) *Contested Ground: Public Purpose and Private Interest in the Regulation of Prescription Drugs*. Oxford: Oxford University Press.

Harris, R. (1964) *The Real Voice*. New York: Macmillan.

Inman, W.H.W. (1986) *Monitoring for Drug Safety*. Lancaster: MTP Press.

Irwin, A. (1985) *Risk and the Control of Technology: Public Policies for Road Traffic Safety in Britain and the United States*. Manchester: Manchester University Press.

Jasanoff, S. (1990) *The Fifth Branch: Science Advisers as Policy-makers*. Cambridge, MA: Harvard University Press.

Jeffereys, D. (1993) Experience of the MCA with multi-state applications, *Drug Information Journal*, 27, 33–8.

Latour, B. and Woolgar, S. (1986) *Laboratory Life: the Construction of Scientific Facts*, 2nd Edition. New Jersey: Princeton University Press.

Lexchin, J. (1990) Drug makers and drug regulators: too close for comfort, *Social Science and Medicine*, 31, 1257–63.

Litchfield, J.T. (1961) Forecasting drug effects in man from studies in laboratory animals, *Journal of the American Medical Association*, 8, 33–8.

Lumley, C.E. and Walker, S.R. (1985) A toxicology databank based on animal safety evaluation studies of pharmaceutical compounds, *Human Toxicology*, 4, 447–60.

Lumley, C.E., Parkinson, C. and Walker, S.R. (1993) The value of the dog in long-term toxicity studies: the CMR international toxicology database, *Adverse Drug Reaction Toxicology Review*, 12, 543–61.
Mann, R.D. (1987) *Adverse Drug Reactions: the Scale and Nature of the Problem and the Way Forward*. Carnforth, Lancs: Parthenon Publishing.
Medawar, C. (1992) *Power and Dependence: Social Audit on the Safety of Medicines*. London: Social Audit.
Medicines Control Agency (MCA) (1991) *MAIL*, 68.
Medicines Control Agency (MCA) (1993) *Towards Safe Medicines*. London: HMSO.
Middlemas, K. (1979) *Politics in Industrial Society: the Experience of the British System since 1911*. London: Andre Deutsch.
Millstone, E. (1978) A framework for the sociology of knowledge, *Social Studies of Science*, 8, 111–25.
Ministry of Health/Scottish Home and Health Department (1963) *Safety of Drugs: Final Report of the Joint SubCommittee of the Standing Medical Advisory Committees*. London: HMSO.
Mitnick, B.M. (1980) *The Political Economy of Regulation*. New York: Columbia University Press.
Parkinson, C. (1992) Twelve month non-rodent studies debated at international forum, *CMR News*, 10, 1–3.
Personal Communication (1995) Telephone interview with Dr Berridge, preclinical assessor at the MCA, Department of Health, London.
Pharmaceutical Society of Great Britain (PSGB) (1962) 'Magnificent' export performance, *Pharmaceutical Journal*, 3 November, 445.
PSGB (1963a) The industry and the Health Service, *Pharmaceutical Journal*, 4 May, 417–18.
PSGB (1963b) Committee on Safety of Drugs: members and terms of reference, *Pharmaceutical Journal*, 8 June, 534.
PSGB (1963c) Committee on Safety of Drugs: memo to manufacturers and importers, *Pharmaceutical Journal*, 26 October, 433.
PSGB (1966) Safety of Drugs: Dunlop Committee Second Report, *Pharmaceutical Journal*, 23 July, 86–7.
PSGB (1967a) Safety of Drugs: Committee's Annual Report, *Pharmaceutical Journal*, 15 July, 59–60.
PSGB (1967b) The White Paper: comments by ABPI, *Pharmaceutical Journal*, 16 September, 246–7.
PSGB (1968a) Industry, safety, Sainsbury and the Bill, *Pharmaceutical Journal*, 9 March, 274–5.
PSGB (1968b) Medicines Bill receives cautious approval, *Pharmaceutical Journal*, 24 February, 215–18.
PSGB (1976) CSM asked to speed up procedures, *Pharmaceutical Journal*, 17 July, 46.
Review Panel of New Drug Regulation (1976) *Assessment of the Commissioner's Report of October 1975 (Summary)*. Washington, DC: US Government Publications Office.
Review Panel of New Drug Regulation (1977) *Summary of the Special Counsel's Conclusions*. Washington, DC: US Government Publications Office.

Salsburg, D. (1983) The lifetime feeding study in mice and rats – an examination of its validity as a bioassay for human carcinogens, *Fundamental and Applied Toxicology*, 3, 63–8.

Schwartzmann, D. (1976) *Innovation in the Pharmaceutical Industry*. Baltimore: Johns Hopkins University.

SCRIP (1988a) *World Pharmaceutical News*, 1270, 24.

SCRIP (1988b) *World Pharmaceutical News*, 1279, 3.

SCRIP (1991a) *World Pharmaceutical News*, 1595, 8–9.

SCRIP (1991b) *World Pharmaceutical News*, 1635, 2.

Silverman, M. and Lee, P.R. (1974) *Pills, Profits and Politics*. Berkeley: University of California Press.

Smart, R.D. (1981) Foreword, *ABPI Annual Report 1980–81*. London: ABPI.

Steward, F. and Wibberley, G. (1980) Drug innovation – what's slowing it down?, *Nature*, 284, 119.

Temin, P. (1980) *Taking Your Medicine: Drug Regulation in the United States*. Cambridge, MA: Harvard University Press.

US Congress (1969) *Drug Efficacy (Part 2)*. Hearings before a Subcommittee of the Committee on Government Operations House of Representatives. Washington, DC: US Government Publications Office.

US Congress (1970) *The British Drug Safety System*. Twenty-second Report Committee on Government Operations. Washington, DC: US GPO.

US Congress (1974) *Examination of the Pharmaceutical Industry (Part 7)*. Joint Hearings before the Subcommittee on Health of the Committee on Labour and Public Welfare and the Subcommittee on Administrative Practice and Procedure of the Senate Committee on the Judiciary. Washington, DC: US Government Publications Office.

US Congress (1982) *The Regulation of New Drugs by the FDA*. Hearings before a Subcommittee of the Committee on Government Operations House of Representatives. Washington, DC: US Government Publications Office.

US Congress (1983) *FDA's Regulation of Zomax*. Hearings before a Subcommittee of the Committee on Government Operations House of Representatives. Washington, DC: US Government Publications Office.

US Congress Process (1986) *Oversight of the New Drug Review Process and FDA's Regulation of Merital*. Hearings before a Subcommittee of the Committee on Government Operations House of Representatives. Washington, DC: US Government Publications Office.

US Congress (1987) *FDA's Regulation of the New Drug Suprol*. Hearings before a Subcommittee of the Committee on Government Operations House of Representatives. Washington, DC: US Government Publications Office.

US Congress (1988) *FDA's Regulation of the New Drug Versed*. Hearings before a Subcommittee of the Committee on Government Operations House of Representatives. Washington, DC: US Government Publications Office.

Vogel, D. (1986) *National Styles of Regulation: Environmental Policy in Great Britain and the United States*. Ithaca and London: Cornell University Press.

Wade, O.L. (1983) Achievements, problems and limitations of regulatory bodies. In Farrell (ed) *Medicines Review Worldwide – A Patient Benefit or a Regulatory Burden? Proceedings of the Fifth Annual Symposium of the British Institute of Regulatory Affairs*. London: BIRA.

Walker, M.J. (1993) *Dirty Medicine: Science, Big Business and the Assault on Natural Health Care*. London: Slingshot Publications.
Walker, S.R. and Lumley, C.E. (1987) Reporting and under-reporting. In Mann, R.D. (ed) *Adverse Drug Reactions: the Scale and Nature of the Problem and the Way Forward*. Carnforth, Lancs: Parthenon Publishing.
Wheeler, D.E. (1963) The President's statement, *ABPI Annual Report*, 5–6.
Wheeler, D.E. (1964) President's statement, *ABPI Annual Report*, 2.
Wilson, G.M. (1962) Assessing new drugs, *New Scientist*, 297, 26 July, 196.
World Health Organisation (WHO) (1969) Principles for the testing and evaluation of drugs for carcinogenicity, *Technical Report Series*, 426, 15.
WHO (1974) Assessment of the carcinogenicity and mutagenicity of chemicals, *Technical Report Series*, 546, 8–9.
Zbinden, G. (1987) *Predictive value of animal studies in toxicology*, Centre for Medicines Research Lecture. Carshalton, England.

7. 'Strange bedfellows' in the laboratory of the NHS? An analysis of the new science of health technology assessment in the United Kingdom

Alex Faulkner

Introduction

A recent disease-focused history of the development of medical research in the United Kingdom, written by a professor of medicine, refers in passing to a new type of health care research. The author notes the multidisciplinary nature of this type of 'medical research' and draws our attention to alleged difficulties of research methodology:

> This new branch of medical research brings together some strange bedfellows, ranging from psychology and the social sciences to biomathematics. It presents many difficulties, not least the uneasy amalgamation of the relatively 'soft' science of interviewing techniques with some fairly sophisticated mathematics (Weatherall 1995: 312).

Current developments in the National Health Service suggest that research knowledge is seen amongst state authorities in the United Kingdom as a means of improving and controlling the development of healthcare services. The new type of health care research known as Health Technology Assessment (HTA) has emerged as both the largest financially and, symbolically, the highest-profile of the research programmes within the new NHS R&D strategy (Department of Health 1993). The NHS Executive has spearheaded the state's co-ordinating action in this area, in spite of the existence of other agencies with kindred health policy interests in the Department of Health. HTA is a multi-faceted movement constituted in the interactions of state policymaking bodies, the medical and healthcare professions, academia, hospital management and healthcare commissioning authorities. Overall the NHS R&D strategy aims to increase the proportion of annual NHS expenditure on research and development from 0.9 per cent to 1.5 per cent. It explicitly advocates the development of an evaluative culture within the NHS, aimed at developing a 'research-based' or 'knowledge-based NHS' (Department of Health 1993). The 'exploitation' of knowledge as a resource generated through research is one of the most important requirements of contemporary industrial capital (Webster 1994). New structures and management in public sector services are accompanied by expansion of research and other knowledge generating practices (Hoggett 1991, Hughes and McGuire 1992). A consequence of this is the formation of new strategic alliances between the various producers and consumers of research knowledge, in which research disciplines participate in the strategic action of healthcare policymaking agencies.

The scope of the national health technology assessment movement brings into view many healthcare issues of interest to a sociological analysis. It raises issues to do with interests, values and inter-organisational relationships. It raises questions about the shaping of research agendas, healthcare policies, regulation of health technologies and, indeed, about the evolving patterns and methods of healthcare which we might have to call upon as health service users. It also raises issues germane to some of the prime concerns of the sociologies of medicine/healthcare and science/technology, including trust and contestability in medical authority, construction of health and healthcare risks, rhetorics of scientific projects and knowledge claims, relationships between the disciplines of medical and healthcare knowledge, and relationships between healthcare experiments, laboratories and technology tests.

In this chapter I am concerned with only a limited set of the possible themes. The focus, therefore, is upon the use of rhetorical discourse in the construction and shaping of a 'need' for health technology assessment around new constellations of institutions and disciplines – 'strange bedfellows' – which are negotiating agendas in healthcare knowledge. I locate this analysis by developing the notion of the NHS as a massive laboratory in and around which healthcare knowledge is produced. Implications for key themes in the sociological study of science/technology and health/illness are discussed. I confine myself largely to considering some of the key activities and developments which are part of the formal national NHS HTA movement in the United Kingdom, with some reference to related activities in the Medical Research Council (MRC), rather than the wider range of healthcare research activity much of which might also follow a broadly health technology assessment model.

Before moving on to consider the main themes of the chapter, I sketch briefly the state-co-ordinated formal structure which has been created for national HTA activity in the United Kingdom, followed by a description of the methods used in producing the analysis of HTA presented here.

Formal organisation and function of national HTA

Health technology assessment is the only one of the many research programmes, set up under the new NHS Research and Development Directorate, to receive support through the formation of a permanent standing group to oversee it, the Standing Group on Health Technology (SGHT). The SGHT identifies priorities for assessment through nation-wide consultation, inviting topic suggestions from health professionals, representative bodies and others. Suggestions are considered by six advisory panels designed to reflect the full range of different sectors within healthcare. Five panels deal with healthcare sectors or types of technologies: the acute sector, primary and community care, pharmaceutical, diagnostics and imaging, and population screening panels. The sixth is concerned with the methodology of HTA. These groups score and rank technologies suggested for assessment which are then passed to the SGHT. SGHT performs a similar exercise to reach recommendations for commissioning assessment projects from research organisations. Membership of these panels and the standing group are of importance when considering the constituencies represented in national HTA. The major criteria for assigning priorities to assessments are stated to be: benefits in terms of improved outcomes for patients; methodological gains; timescale of potential benefits; value for money of assessment; importance of *early* assessment; and factors relating to Health of the Nation policy, prevalence and social/ethical considerations (Department of Health 1995: 46).

Methodology

For data about HTA in the United Kingdom, I draw upon reports published by the major new institutions representing the HTA movement, minutes and papers relating to meetings of advisory groups, comment and debate in the medical and health services press, discussion and communications with some key participants, and upon my own participation in HTA activity, akin to the 'participant comprehension' which Collins (1984) has described. Collins assumed that the researcher adopting this approach would have a single, clear professional identity as sociologist, but nevertheless describes how he and colleagues 'became scientists ourselves' (1984: 60). My professional affiliations are both to sociology and 'health services research'. For one year, in the capacity of a health services researcher, I became a 'scientific secretary' to one of the six advisory groups working with the major group charged with co-ordinating national HTA, the Central Research and Development Committee's Standing Group on Health Technology. I have also been involved in academic research work much of which can be described as health technology assessment, and some of which forms part of the national HTA programme. I thus also reflect upon familiarity with the practices, discourses, networks and institutions of HTA to inform the account presented here. Particular 'data' drawing on this experience are presented in the text in quotation marks and noted parenthetically as being 'author observation'.

Laboratories, experiments, technology testing and rhetoric

There have been recent calls for and signs of a rapprochement between the sociologies of medicine/healthcare and science/technology (Bartley 1990, Berg 1995, Casper and Berg 1995). The notion of a 'knowledge-based NHS' is of dual interest because it signals a bringing-together of scientific knowledge, which has been investigated in social studies of science, with healthcare practice, which has been investigated by sociologists of medicine and healthcare. Central to the practice of science are laboratories. Recent developments in the sociology of scientific knowledge and the history of science have regarded laboratories as both empirical (*e.g.* Latour and Woolgar 1979) and metaphorical (*e.g.* Macleod and Rehbock 1994) locations for the study of scientific activity and the production of scientific knowledge. Bartley (1990) suggested that medical sociology, using the tools and concepts of the sociology of science, might investigate medical scientists' empirical laboratories as the locations where medical knowledge is constructed. The notion of the NHS itself as a massive laboratory enables such approaches to be applied to the broader fields of healthcare practice and policy which come under scrutiny in the HTA enterprise.

The primary form of scientific activity which takes place in laboratories is the experiment and its interpretation. In social studies of science the variable relationship between experiments and laboratories is an important subject (Knorr Cetina 1992): ' . . . laboratories and experiments combine differently in different fields' (1992: 114). The sociological focus upon laboratories has permitted a view of science which goes beyond taken-for-granted notions about the ability of experimental methodology to support or negate hypotheses by applying bias-eliminating designs, to allow consideration of experiments in the context of the resources and practices employed in conducting them. Laboratories are not merely the location in which experiments are conducted, but they also involve, for example, the deployment of equipment and measurement instrumentation, the formation of strategic inter-individual and inter-organisational alliances, and the use of persuasive literary techniques in the presentation of 'findings' of experiments in scientific publications.

MacKenzie (1989) has drawn attention to the similarity between scientific experiment and technological testing. The approaches developed for the understanding of scientific knowledge can be applied also to the examination of the testing of (hardware) technologies. While the construction of scientific knowledge typically involves interpretation of particular observations in terms of theory, technology testing involves such interpretation in terms of predicted 'real-world' performance. Knorr Cetina (1992: 116) has described the laboratory as an 'enhanced environment', in which often obscure underlying processes are rendered legible by means of instrumentation and measurement. Thus experimentalists and technology testers work with 'traces' of processes rather than the processes themselves. Technology testing involves 'projection' from the test to observed performance (Pinch 1993). Observations produced 'under laboratory conditions' or under test conditions, are projected by interpretive techniques as predictions of how a technology would perform if applied in the real world. Thus the concepts which have been developed in social studies of science and technological testing can be used in the sociological investigation of the 'assessment', in other words the testing, of medical technologies. Health technology assessments enact various forms of projection, which construct health professionals, healthcare technologies and ourselves (as patients) as participants in the experimental laboratory of the NHS.

The sociology of scientific knowledge (SSK) has also drawn upon social constructionism to show how scientific facts are constructed through the manifold practices of scientists (Latour 1987). Key to this approach to science has been the concept of rhetoric. Having been historically regarded as antithetical to science, rhetorical discourse is seen in SSK as an intrinsic aspect of the methods used by scientists not merely in publicising the results of science but in the constitution of scientific theory and fact (Beer and Martins 1990). Rhetoric in this context can be defined as discourse which implicitly or explicitly persuades or proposes. Here the term has no pejorative connotation. Disciplines have disciplinary rhetorics, and fields of science characterised by the activity of several disciplines, such as health technology assessment, are likely to have multi-disciplinary or inter-disciplinary rhetorics. We can thus speak of the rhetorical constitution of disciplines or scientific fields as an aspect of the production of scientific knowledge.

The deployment of disciplinary rhetorics constitutes 'boundary-work' in Gieryn's sense (Gieryn 1983). In this conception, disciplinary discourse is to be seen as laying claim to professional territory by defining appropriate methods, concepts and agendas for a scientific programme. Such rhetorical work may be especially characteristic of new, changing, weak or interdisciplinary fields of scientific activity, where shared meanings and concepts are lacking (Porter 1995: 228). Tension between disciplinarity and interdisciplinarity is likely (Good 1993). Health technology assessment in the United Kingdom is just such a field of scientific activity.

The relation of disciplinary knowledge to the NHS is a matter which disciplinary activists play a part in constructing. This has been shown, for example, in the work of Ashmore *et al.* (1989) in their discussion of health economics in relation to the medical profession. The authors regard health economists as engaged in an 'educative strategy' (1989: 186) using programmatic discourses, in other words, discourses describing and promoting programmes of desirable change in NHS practices, which might be achieved by the adoption of lessons from the discipline of economics. Similarly, Pinch *et al.* (1992) discerned rhetorical devices at work in the social processes inherent in the introduction of clinical budgeting technology in the NHS. In this chapter I focus primarily upon 'formal' rhetoric produced in policy statements and 'public' professional scientific debate to analyse the project of national HTA.

Constructing and supporting a legitimate 'need' for health technology assessment

In the late 1970s and early 1980s, a repertoire of accounts emerged in the discourse of international healthcare policymaking which might be drawn upon by different actors in constructing rhetorical justifications of a need for HTA. The first strand in this repertoire is a construction of policymakers' 'concerns'. In the late 1970s, a concern to promote cost containment as the key issue was paramount amongst policymakers' discourse on medical technology. For example:

> Technology has been identified as a major cause of increasing health care expenditures, . . . controlling new technology is required to contain health care costs (Committee on Technology and Health Care *et al.* 1979: 14).

Such accounts referred frequently to the uncontrolled proliferation of 'new medical technology', especially novel, high-cost, high-tech diagnostic equipment such as the CT scanner and magnetic resonance imaging (Jennett 1986). As the above account indicates, part of the policymakers' agenda was to control diffusion of this type of equipment because it caused escalating expenditure. In the perspective adopted here, this 'cost-containment' justification of a need for HTA is regarded as a rhetorical device which is constitutive of the boundaries of HTA, enabling certain interests and disciplines to lay claim to a stake in its activities. It does not follow that financial expenditure is the aspect of 'costs' which will necessarily be dominant in the evolving model of HTA practice. While concern for expenditure might have been presented as the primary cause, a *set* of 'concerns' has now come to be portrayed in healthcare policymakers' discourse as underlying the need for assessment of technologies:

> There is growing concern relating to the health benefits and risks of technology, its financial costs, and its social implications (Banta and Gelijns 1987: 255).

The first of these authors is a high-profile international adviser on health technology policy who was involved in meetings such as a seminal and symbolic multi-interest 'Tidal Wave' conference held in England in 1991 (Hoare 1992). The set of concerns Banta cites is notable for its reference not only to health benefits and risks, but also to social implications, a theme which can be traced back to the birth of the original technology assessment movement in the United States, where the concept emerged in the mid-1960s associated with liberal political movements. Prominent in this early concept were notions of the unintended consequences and indirect effects of technologies such as industrial processing (O'Brien and Marchand 1982: 7). This model has been taken up within the programmatic discourse on HTA in the United Kingdom. Another high-profile expert and early proponent of assessment of medical technologies, Barbara Stocking, conducted case studies of expensive health-care technologies in the United Kingdom. Then Director of the King's Fund Centre for Health Services Development, one of the United Kingdom's major independent research centres for research and consultancy on health services issues, she described technology assessment as including:

> the technical and clinical evaluation of a technology, as well as its economic, social and ethical implications (cited in Hoare, 1991: 1).

Rhetorically deployed concerns about these different dimensions of technology support a 'need' for HTA activities in the NHS. The grouping of 'concerns' of healthcare policymakers also promotes a particular model of the disciplines which might appropriately participate in it. In an article in the *Lancet* aimed at a medical audience, the current (1997) Director of NHS R&D explicitly makes a case for combining specialist disciplines:

> The need to reassemble these fragments and to define their limits is central to the successful incorporation of science into health care (Swales 1997: 1319).

He identifies the major topics of this science as being medical outcomes, cost and quality of life (1997: 1320). The second rhetorical strand is thus a programme to amalgamate diverse disciplines: multi-disciplinarity. Linked to multi-disciplinarity is the 'discovery' of common ground between disciplines via the dusting down and applauding of new 'founding fathers'. This is seen in the canonisation of individuals who according to HTA activists espoused and promoted experimentalism and the monitoring of the effects on patients of healthcare practitioners' work. Individuals who are being acclaimed in this rhetorical discourse include, notably, the epidemiologist Archibald Cochrane who is discussed below in relation to the role of epidemiology and methodology in HTA. For example, Cochrane has been commemorated recently by the publication of a collection of essays in his honour (Maynard and Chalmers 1997), and 'a large photographic portrait of Cochrane has pride of place in the reception area of our (academic) department' (author observation). Two other individuals have been elevated in this way. Ernest Codman, an American who in 1900 instigated a hospital system for monitoring the end results of surgical care, was praised by the Director of the NHS Centre for Reviews and Dissemination (Sheldon and Faulkner 1996). And the 18th-century surgeon John Hunter, credited within the surgical profession as being the first scientific surgeon, received a strikingly titled homage from the chairman of the SGHT (Irving 1993).

Appeals to common ground can be seen in the heart of the state-promoted definition of 'health technology' itself. What counts as technology? The meaning of 'technology' is indexical (Pinch *et al.* 1992), that is, it is related to the occasions and contexts of its use. It is also a metaphor. Some metaphors become more dominant than others, and thus can be interpreted as traces of the workings of power across communities, networks or disciplines (Leigh Star 1991: 52). Whereas earlier policymakers' discourse concerned expensive new medical equipment, health technology has been promoted in the programmatic discourse of the state-orchestrated HTA movement in the United Kingdom as a highly abstract, all-embracing metaphor. Far from being confined to high-cost high-technology equipment, it has been promoted by the NHS Executive (and this was a usage already common amongst healthcare policymakers in a number of other countries) as being:

> . . . deliberately defined as broadly as possible. It encompasses all methods used by health professionals to promote health, prevent and treat disease and improve rehabilitation and long term care. It includes the activities of the full range of health care professionals, the use of equipment and procedures, and the administration of pharmaceutical products (Department of Health 1995:8).

Health technology is the experimental matter in the laboratory of the NHS. It forms part of the discourse of the 'knowledge-based NHS'. The boundaries of health technology are being drawn very broadly, by the NHS Executive arm of the state.

It is possible to characterise in a number of ways the range of topics which fall within the purview of health technology assessment. In order to draw attention to the metaphorical and cross-disciplinary deployment of health technology terminology, I present below a typology which characterises technologies from the viewpoint of a healthcare practitioner, with examples mostly drawn from the HTA programme.

A typology of 'health technologies' from a health professional's perspective

Type of Technology	*Examples*
Information systems	Picture Archiving and Communication System (PACS)
Material artefacts	Drugs, bronchodilator, hearing aid
Organisations	Regionalisation (of intensive care), specialised versus local access service (for vascular surgery), primary care-based emergency centre
Interpreted techniques	Diagnostic technologies and screening tests
Technique-assisted interventions	Physiotherapy, laser treatment, coronary artery bypass graft, knee prosthesis
Interpersonal communication	Counselling, psychological treatment, health visitor domiciliary visiting

For Pinch *et al.* (1992), health technologies are 'social technologies' in that they are intended to change human behaviour. This notion can be extended by asking what sorts of behaviour might be changed. Here the targets of intended change are health professionals and healthcare policies. Examples would be modes of organisation of healthcare delivery, healthcare techniques, or practitioners' choice between alternative healthcare options. The majority of HTA topics focus upon a condition (such as stroke or low back pain) and its corresponding technologies (such as rehabilitation techniques or spinal surgery) (Department of Health 1995). Some topics include questions of organisational models for the utilisation of technologies, division of labour, job design, and skills. The National Co-ordinating Centre for HTA is charged with developing an inventory of existing health technologies and briefings on their effectiveness and cost-effectiveness (Research and Development Directorate 1995). This re-interpretation of the earlier principles of HTA is commensurate with the radical approach to creating a 'knowledge-based' health service which the state has promoted. The move from new hardware-embodied technologies to a more general notion of interventions having health effects is again evident.

Further rhetorical discourses take the form of 'horror stories'. First, identifiable primarily with the health care purchaser's perspective,[1] it has been a frequently quoted statistic that only some 15–20 per cent of health care interventions had a firm basis in research evidence (Hoare 1992). The editor of the *British Medical Journal* greeted the launch of the first annual report of the Standing Group on Health Technology with the words: 'Few decisions made in health services are made with good evidence' (Smith 1994). So a vast array of healthcare practices are being performed, in this account, without the 'evidence' to 'support' them. This has been countered by proponents of 'evidence-based medicine' amongst the medical professions who uphold the need for individual clinical expertise alongside a commitment to draw on the best available evidence in making care decisions (Sackett *et al.* 1996). This counter claim can be viewed as a version of the conventional medical professional strategy of appeals to a need for clinical freedom.

The second type of horror story takes the following form: unevaluated health technologies can be dangerous and a risk to health. This is illustrated by the construction of the need for assessment of health technologies by reference to previous health technologies which, allegedly unassessed, found their way into routine medical practice only for their deficiencies then to become apparent. The classic example, frequently quoted, is that of gastric freezing (Challah and Mays 1986, Department of Health 1995: Foreword), in retrospect a painful and ineffective treatment for duodenal peptic ulcers.[2]

There are thus several linked strands in the rhetorical repertoire at the disposal of proponents of HTA. A rhetoric of broad rational concerns, employing appeals to medical, social and economic values is linked to appeals to cross-disciplinary common ground; this is supported by a rhetoric of appeal to possible danger and risks from healthcare provided in the absence of an adequate evaluative knowledge base. These appeals are not specific to particular disciplines within HTA, and so they promote multi-disciplinary participation in it. Appeals are both to clinical and economic concerns and to a need to produce new healthcare knowledge. Health technology is a 'trans-disciplinary' concept, like other such notions 'metaphorically encompassing the several parts of material handled separately by specialised disciplines' (Good and Roberts 1993: 6). However, as the above interpretations suggest, these rhetorical resources are flexible in their potential use, and are contestable. This draws our attention to power relations at work in the negotiation by the various disciplines for positions in the multidisciplinary world of HTA.

Consideration of these persuasive discourses would be important in a sociological history of the emergence of HTA in the United Kingdom. Alongside them should be set an historical account of the 'official' policy development which identified a need for nationally co-ordinated HTA. This of course would be another, but different, rhetorical discourse. It is thus worth noting the appearance during the 1980s and more recently of committees and advisory groups, and reports from national bodies, which would doubtless form part of such an account. These include: a call from the Council for Science and Society in 1983 for evaluation of all expensive new medical technology; a Medical Research Council (MRC) committee dealing with research on health service delivery; a national Health Technology Assessment committee established in the Office of the Chief Scientist in 1987; a House of Lords Select Committee on Science and Technology which reported on 'Priorities in Medical Research' in 1988, noting a wide disparity between the research needs of the NHS and the activities of the MRC; and a government advisory committee reporting on 'Medical Research and Health' in 1993, explicitly supporting the concept of health technology assessment.

Research alliances and institutional co-ordination

Via HTA, the NHS is gaining access to the means of production of research knowledge, and knowledge-producing disciplines and institutions are gaining access to the laboratory of the NHS. Research alliances are emerging in interactions and negotiations between state orchestration, healthcare organisations and the disciplines of healthcare knowledge production. It is not possible here to analyse the full range of processes which constitute the organisational and disciplinary membership of HTA. In this section, therefore, I give examples of alliance-formation, co-ordination and centralisation which are evident. In doing so, I am aware that I am glossing over many of the tensions which undoubtedly exist around the state-promoted HTA project.

The participating constituencies in HTA are broad. The ways in which individuals and institutions become engaged in HTA networks include memberships of the SGHT and its advisory panels, location of co-ordinating bodies such as the National Co-ordinating Centre for Health Technology Assessment within academic institutions, involvement of 'acknowledged experts', often researcher-clinicians, in the process of defining assessment questions prior to funding, consultation within the NHS and more widely amongst representative groups such as medical royal colleges and the Patients' Association, and the advertising and award of assessment project funds. The SGHT and its advisory panels have multi-disciplinary membership, including healthcare purchasers and providers, clinicians and other health professionals, health economists, and individuals seen by members of the R&D Directorate as methodology specialists.

Closer alliances between academic researchers (from the disciplines of epidemiology, medical statistics, public health medicine, human sciences, psychology, sociology, social anthropology and economics), clinician-researchers and purchasing authorities (especially in the form of public health departments) are being developed. Novel forms of collaboration are arising between academic researchers and health service practitioners. Examples are the location of health service research centres in NHS hospitals, and formal collaborations between NHS public health departments and academic research centres.

At inter-institutional level, long-standing co-ordination between the MRC and the Health departments has been extended (Medical Research Council 1991). In 1992, health services research funded by the MRC was seen to need its own managerial Board, the Health Services and Public Health Research Board (HS&PHR) (Medical Research Council 1992, Medical Research Council 1995). The research councils, the Department of Health, the MRC HS&PHR Board and the SGHT are all linked by cross-memberships.

Thus there are multiple signs of new inter-institutional networks, alliances, co-ordination, and centralisation in the broad institutional development of national health technology assessment. The growth of co-ordination of health technology assessment activity attests to the orchestrating role being played by state agencies. However, the term 'health technology' has not been strongly embraced by academic producers of HTA knowledge – 'they generally prefer to describe their organisations, as we do, in different terms, such as "health services research" or "health care evaluation" or "health science"' (author observation). This terminology can be interpreted as signalling an independent position in preserving a capacity for determining local research strategies without becoming over-dependent upon the NHS-driven agenda of HTA.

Given these organisational developments, the following section considers the relationships between the disciplines and methodologies of HTA.

Multidisciplinarity, experimentalism and accuracy of projection

In health technology assessment, projection – the extrapolation of findings from the test environment to the real world – takes a number of forms depending on the methods adopted by the different disciplines of healthcare knowledge. The current incarnation of HTA in the United Kingdom comprises a multi-disciplinary approach drawing upon both quantitative and qualitative methodologies, with a distinctive mixture of disciplines and skills. The research commissioning groups of the HTA programme and the MRC have a preference for research teams made up of certain disciplines. Informal discussion with HTA administrative support staff confirms that projects lacking clinical specialists, elements of economic evaluation or access to expertise in medical statistics are unlikely to be supported. In seeking tenders for a National Co-ordinating Centre to manage and support national HTA activity, the NHS Executive sought a single centre capable of providing: 'Multidisciplinary scientific skills . . . [which] should include clinical (medical, nursing or therapy) and epidemiology, health economics, sociology and other social science disciplines' (Research and Development Directorate 1995).

As Pickstone (1993) argues, until very recently the experimentalism of basic biomedical research has been the primary world of medical research, clinical medicine a 'poor, confused imitation' (1993: 452), and research has been discipline-driven. The impact of scientific research on health service delivery was relatively indirect and long-term (Austoker 1989, Booth 1989, Medical Research Council 1995a). With the arrival of HTA, the organisational changes described above have been implicated in a re-alignment and shifting of power across and between professional disciplines, and a disruption of the previously strong boundary between high status experimental biomedical science and the scientifically lower status clinical research.

Exponents of epidemiology and medical statistics have been the leading disciplinary activists in linking experimental science and clinical practice. The primary form of projection ('generalisability' in the language of HTA), promoted as the vehicle of this linking, is the randomised controlled trial (RCT). In principle the RCT enables valid generalisations to be made because it eliminates sources of bias. In promoting the RCT as the preferred methodology for what he called 'applied medical research', Archibald Cochrane – in what proponents of HTA now regard as a core text in the application of the experimental method to health service research (Cochrane 1971) – aligned clinical research with the experimentalism of the more prestigious biomedical sciences. Skills in experimental research design for causal analysis of disease patterns are strongly claimed within the disciplinary boundaries of epidemiology. Its practitioners have advanced their influence, applying the same methodological concepts to questions of technological cause and health effect within health services. Cochrane's name is now used rhetorically to symbolise an international search for bias-free evidence about the effects of health technologies, gained by the collection of results *only* from RCTs. This is known as the Cochrane Collaboration.

Further entrepreneurialism in the discourse of epidemiology can be seen in the 'systematic review', a second methodology aimed at identifying bias-free, generalisable relationships between health technologies and their effects. In strong programmatic statements from policy centres for HTA, this assessment method is also construed as a scientific method owing its principles to epidemiology. The method has been enshrined in the Cochrane Collaboration handbook on 'Preparing and maintaining systematic reviews' (The Cochrane Collaboration 1994). Its introduction, entitled 'The science of reviewing research', begins:

> The scientific principles that apply to epidemiological surveys apply also to systematic reviews: a question must be posed, a target population of information sources identified and accessed, appropriate information obtained from that population in an unbiased fashion, and conclusions derived. Often statistical analysis can help in reaching conclusions (1994: VI-1).

Here the information sources referred to are the results of existing, completed RCTs. The epitome of the method, referred to above as 'statistical analysis', is meta-analysis, the statistical summarisation of the data from several empirical studies deemed to be comparable (Dickersin and Berlin 1992). The formal commitment to elimination of bias has resulted in an attempt to construct the systematic review also as an experimental scientific procedure, along the lines of the randomised controlled trial.

However, a purely quantitative approach to systematic reviews is implicitly criticised by the NHS' national centre for this activity in the formal guidance it has issued (NHS Centre for Reviews and Dissemination 1996). It commends an approach which:

> considers all the results taking into account not only the methodological rigour, and therefore reliability of these studies, but also helping to highlight and explore differences. A qualitative analysis of the evidence is therefore an essential step in the assessment of the effectiveness of a health technology (1996: 46).

The document bemoans an exclusive focus upon the 'narrow' methods of statistical pooling of data. These passages suggest that rather than being a matter – for those with appropriate statistical expertise – of cumulation of results from separate studies, the review requires qualitative judgements to be made about the homogeneity, or otherwise, or studies. Thus what is at issue is the basis on which accurate projections from 'results' of experiment (or combined experiments) might be made: in other words of how bias or subjectivity might best be controlled.

Thus the contribution of epidemiological discourse to the construction of the HTA agenda powerfully promotes scientific experimental methodology. However, even around this formal programmatic agenda, conflict is in evidence, for example over the status of the RCT (Black 1996) and of meta-analysis (Eysenck 1994) as methodologies of knowledge production.

Epidemiology also figures in the framing of the national HTA agenda through its focus upon the prevalence of medical conditions. Whereas historically clinical science has been greatly occupied with relatively rare conditions (Frankel 1989), epidemiological analysis of the distribution of remediable conditions has paved the way for a re-distribution of clinical scientific research attention to some of the most common afflictions. And, as noted earlier, prevalence is one of the criteria explicitly used in the formalised priority-setting processes of the SGHT in identifying topics for assessment.

There is increasing interest on the part of medical practitioners in so-called 'qualitative' research methods, coming from within the arena of clinical research. For example, there has been a series of editorials, appearing in the most prestigious medical journals, including the *British Medical Journal*, presenting the case for qualitative methods in healthcare research, and analysing the interface between qualitative methods and clinical experimental research designs. Programmatic statements from a medical viewpoint shape the relationship between qualitative and experimental research. Qualitative research is held to be especially relevant for hypothesis generation, explanation of experimental or quantitative findings and understanding of factors affecting implementation of research results (Jones 1995). Papers written by sociologists have also been published in medical journals, outlining some of the main qualitative research methods which can be applied in healthcare research (*e.g.* Mays and Pope 1995).

The definition of medical symptoms and outcomes of treatment for research purposes has, historically, been largely the preserve of medical specialists performing clinical research. In the new model of health technology assessment, the boundaries of this territory appear less clear. New contributions are being made predominantly by research techniques derived from sociology and psychology. This is embodied in the increasing development of measures of health status and health 'outcomes' measures, in particular measures informed by our own experience, as health service users, of our symptoms and health states. Patient-derived measures of the 'outcomes' of 'interventions' – are being increasingly incorporated into the array of measures used in experimental tests of health technologies. The novelty is that they are constructed from 'qualitative' research, usually interviews, into our experience of symptoms. A typical example is the development of an 'instrument' to assess the outcomes of total hip replacement surgery – one of the highest priority technologies identified in the early HTA programme – from patients' perspectives: 'Many questionnaires . . . intended for general use . . .

may be . . . insensitive to the specific changes in health produced by a particular intervention . . . Questionnaires are needed therefore which address patients' perception of a single disease entity . . . ' (Dawson *et al.* 1996: 185). These authors' disciplines combine sociology and orthopaedic surgery. Psychometric statistical methods in particular are being used to 'validate' such measures. Validation in this sense is another form of 'projection', akin to the testing of measurement instrumentation technology in laboratory settings. The construction and validation of such measures are presented as processes establishing the scientific accuracy of the representation of patients' subjective experience. This enterprise results in calls for the 'standardisation' of such measures (McDowell and Jenkinson 1996), because the generalisability of results depends partly upon having reliable methods of enabling the effects of the application of health technologies to produce 'traces', to use the SSK term described above, which can be apprehended, quantified and interpreted by statistical analysis.

The proliferation of measures of health status and quality of life extends the definition of what counts as an effect of clinical intervention, and thereby re-shapes the meanings which health and illness might have in the discourse of healthcare policymaking. This process is being strongly supported by the involvement of proponents of health economics in HTA. Health economics has established itself as a core discipline in the health technology assessment repertoire. Economic evaluations are increasingly a part of clinical trials, and it is policy to encourage this development both in the HTA programme and in the MRC. The performance of economic evaluations alongside clinical trials is promoted by professional economists within the discourse of NHS R&D policymaking as a matter for professional health economists rather than as an activity which can be accomplished by 'doing-it-yourself' (Drummond 1994: Foreword). Assessing costs and benefits in health technology assessments enhances the perceived need for health status measurement tools, because changes in health status, measured either as survival or quality of life, are the 'effects' against which differences in costs between alternative technologies are evaluated in economic analysis. Ashmore *et al.* (1989) have drawn attention to the processes by which economists, in interaction with others, construct quality of life measures (QALYs) as representations of generalised public preferences for states of health. Here it can be said that this too, like the discourse of epidemiology and medical statistics, is a methodology for controlling bias and attempting to take a viewpoint which might somehow make it possible for judgements based on social values to be expressed on behalf of a population. This is explicit in the texts of health economists who state that they perform their analysis 'from the perspective of the NHS' or from the 'societal perspective'. Utilitarian philosophies are preferred. But here too, within the expert system of health economics, conflict and debate can be found. It has been questioned, for example, whether the focus upon the *health* benefits of health technology is too narrow. Health economics might fail to take account of other benefits or costs which, as citizens rather than as patients, we might derive from healthcare (Ryan and Shackley 1995).

In interdisciplinary fields, expertise from more than one discipline comes together to seek common goals (Good and Roberts 1993). Within the constellation of HTA disciplines, the clinical sciences and health economics are the ones most clearly delineated. Some interdependencies suggest a blurring of boundaries between the constituent HTA disciplines, reflecting research alliances which have been formed. These include, for example, an interdependence, centred on outcome measurement, between the disciplines promoting expertise in interviewing and questionnaire design and the corresponding clinical specialisms and body systems. A further interdependence may be discerned around cost-effectiveness between the disciplines concerned with health status and quality of life measurement, and health economics. On the other hand, conflicts and tension exist between disciplinary and inter-disciplinary rhetorics of HTA. For example, 'adherents of qualitative methods commonly criticise HTA's emphasis on the randomised controlled trial' (author observation); a public health consultant reports that, in a leading teaching hospital, 'powerful clinical scientists' believe that 'the HTA side is contentious, as there is an obvious link to rationing . . . [though] . . . formalising an HTA process in the NHS Trust would certainly ring the management bell' (Ayres, P. personal communication); and leading voices in the formal programme of health economics criticise the clinical professions' agenda of applying scientific evidence to healthcare practice (Maynard 1997).

The rhetorical claims of health technology assessment constitute the activity of health technology assessment as having the legitimacy of science at a time when novel alignments of professional disciplines are developing, when professional identities and disciplinary authorities exist in tension with each other, and when research communities are being invited to respond to a culture of health service 'needs'. Those involved in the disciplines of HTA perform experiments in and around the laboratory of the NHS. These experiments use measurement instruments to test health technologies. In these tests projections are made from the laboratory world to the 'real-world' of health technology in healthcare practice. The different disciplines of HTA are able to mobilise their technical, methodological and rhetorical resources to different degrees in constructing the HTA agenda.

The assessment agenda

The rhetorics of HTA discussed above might be taken as a guide to the assessment agenda which we would expect to find embodied in the national HTA programme: 'concerns' for benefit and risks to health, for cost and for social implications. We can examine the HTA programme to see what form these concerns might take, shaped by and shaping the organisational and disciplinary processes described above.

Turning to the aspects of health technologies which are typically assessed, it is clear that 'effectiveness' and 'cost-effectiveness' are the two hinges on which the formal NHS HTA world-view swings. Strong statements confirming this abound: for example, 'The aim . . . is to help those conducting and funding trials to ensure that their work tackles the issues of cost-effectiveness as well as effectiveness whenever it is feasible to do so' (Drummond 1994). The MRC approach is essentially the same. Its studies 'measure effectiveness and efficiency in relation to health outcomes' (Medical Research Council 1995b: 107).

Turning first to costs, in what ways are 'cost implications' being constructed in the HTA agenda? A concern with economic implications might suggest that encouragement of potentially income-generating technologies might be high on the agenda, or that high-technology high-cost equipment would be of priority for assessment, but the voice of commercial exploitation is noticeable by its absence from the formal discourse and practices of national HTA, in spite of policymakers' early plans to create collaboration between the NHS R&D strategy and industry (Department of Health 1993), and attempts by commercial interests to stimulate alliances via national 'research foresight' in health-related technologies.[3] It appears that the strength of the clinical and academic economic discourses on benefits and costs of health technologies have resulted in an institutionally bounded, circumscribed agenda which is relatively impervious to commercial networks and interests. This is not to say that practitioners of HTA research, for example, might not transact with both national HTA and commercial organisations in seeking to negotiate favourable conditions for themselves locally, but the clinico-economic discourse of health costs and benefits effectively excludes commercial interests.

Concerns for risks to health were part of the policymakers' early rhetoric of a need for HTA. Examination of the topics of national HTA suggests that *absolute* safety and risks to health are not the major issues in the HTA agenda. Rather, safety and health risks are aspects of outcome measurement in assessments where different technologies are being compared. Of course, in the case of pharmaceuticals and 'medical devices', statutory agencies exist with responsibility to ensure safety: the Committee on the Safety of Medicines, and the Medical Devices Agency in the Department of Health. But the over-riding focus upon comparative effectiveness and cost-effectiveness in the HTA disciplines projects quite different definitions of risk. The major dichotomy evident here might be characterised, if somewhat glibly, as one between risk to the public health and risk to the public purse. Like processes of risk assessment (Carter 1995), health technology assessments construct boundaries between safety and danger. However, unlike discourses of risk to the healthy body, the discourses of health technology assessment amalgamate concerns with transgression of boundaries of health with concerns regarding transgression of boundaries of financial budgets. They thus generalise and extend the notion of risk. The influence of epidemiological and economic perspectives results in the construction of risk at the level of society as a whole.

In HTA health benefits and costs are aspects of both clinical and economic discourses. The agenda of health status measurement and quality of life measurement has been referred to above. 'Costs' and 'benefits', like 'health technology', themselves develop metaphorical meanings. Discourses of clinical research, incorporating assessments of the effectiveness or efficacy of healthcare technologies derived in interactions with patients, refer to effects of technologies such as complications, mortality, and patient 'outcomes'. Health economics constructs costs and benefits in both financial terms and in terms of the health outcomes of quality of life and mortality. Experimental science, as embodied in the epidemiological and statistical methods of HTA, requires standard, reproducible forms of measurement so that stable comparisons between different patients or groups of patients at different times may be engineered. The other disciplines of HTA deploy techniques which enact this requirement. Economists' cost-effectiveness analyses frequently proceed by constructing a model of a typical health district or general practice population in order to examine cost and benefit implications (*e.g.* Bachmann and Nelson 1996). This represents another form of testing of health technologies, in this case projecting from simulations to actual populations.

Turning to the shaping of 'social implications', it has been argued by leading HTA proponent David Banta that: 'The social implications of a new or existing technology can be the most challenging and difficult aspects of evaluation . . . the methods for assessing social implications are relatively undeveloped . . . ' (Banta and Luce 1993: 132). The implication is that when methods are better developed, HTA will be able to address these issues better. However, it could be argued that if HTA practitioners are to focus on a scientific agenda built around generalisability, elimination of bias, and the representation of a form of aggregated public interest, this would preclude examination of substantive social and ethical issues. Economic evaluation is leading to consideration of ethical issues in healthcare provision, but this is focused upon questions of choice between alternative services (the rationing debate), rather than the examination of the substantive social and ethical issues arising with particular technologies such as, for example, animal to human organ transplantation. 'Xenotransplantation was considered by the Acute Sector Advisory Panel of the SGHT, but not accorded priority, in the first year of operation of the national HTA programme' (author observation). It may be that mechanisms outside health technology assessment, such as the investigative activity of the mass media and specially convened groupings of state-identified experts (Advisory Group on the Ethics of Xenotransplantation 1996), will be the major means by which those technologies, with apparently broad social and ethical implications, are addressed within society. This would be despite the HTA claim to this agenda.

The discourses of epidemiology, statistics and economics construct the population's health status as the indirect aggregate of many individual outcomes of health technology. The current HTA agenda allows *directly* for the patient's voice in closely limited ways. Users' representatives such as the College of Health are included in the consultation process on priority topics for assessment. A few health technology assessment topics include lay or patient attitudes or actions as a major strand in the assessment question. One example is the study of the effect on treatment rates of patient participation in care decisions, using interactive video disks containing personal testimonies of former patients and information about the probability of different outcomes following surgical treatments (NHS Executive 1996). As in this case, projects which do incorporate users' voices contain them within the design of the project. They give priority to measuring the outcomes of that participation by the experimental methods of bias-elimination and in terms of aggregated health outcomes. Multidisciplinary HTA defines the healthcare user population as the object of more or less cost-effective healthcare interventions.

Lay incursion into the expert domains of design, conduct and interpretation of assessments has occurred with AIDS activists in the United States (Epstein 1995). Such engagement is not evident in HTA. And, as Elston (1991: 82–3) has noted, experimental treatments and illnesses where curative medical science has little to offer are specific, limited areas perhaps not representative of the overall capacity of healthcare consumers to challenge the authority of medical science.

The embryonic agenda of HTA, considered at this general level, is not about what the sociology of scientific knowledge is most often concerned with, the processes of building particular facts. It is, rather, about preparing the ground which makes it possible for some sorts of facts – rather than others – to grow, what sorts of facts are going to be important. This is perhaps the reason why so much of the early funding in the national HTA programme has been put into study of the methodology of HTA itself. A reflexive emphasis on methodology reflects and facilitates the engagement of the different professional disciplines 'taking part' in HTA. This is also a debate about what is the appropriate instrumentation for the laboratory of the NHS. It again reflects the tension between disciplinarity and interdisciplinarity in HTA. While the major emphasis has certainly been upon the bias-reduction methodologies of the controlled trial and meta-analysis, no exclusive coalition can be said to have emerged. The voices of non-experimental assessment methodologies and non-population oriented disciplines have also negotiated places – at least for the time being – in the emerging agenda. Tensions are evident, for example, in the small number of studies funded on topics such as the role of qualitative methods and action research in HTA (NHS Executive 1996). 'Similar conflicts occur routinely in informal discussions about the respective status of the randomised controlled trial in comparison to non-experimental assessment methods' (author observation). The effect of such internal ambiguities is to enhance the authority of health technology assessment discourse and its participating voices by retaining control over its agenda. This indeterminacy in turn can be seen as a boundary-constructing control strategy (Jamous and Peloille 1977, Gieryn 1983).

Conclusion

'Health technology' is an abstract metaphor shorn of specific disciplinary or methodological signification. As the quotation at the beginning of this chapter suggests, unresolved tensions between disciplines can exist in the interpretative elasticity which this creates. HTA may or may not be interpreted by different interests as an extension of medical research. The trans-disciplinary concept of health technology has been promoted by the state, and the healthcare knowledge disciplines orient themselves to it in different ways. The metaphor enables these diverse actors including HTA research policymakers and research knowledge producers to shape their activities in diverse ways in and through the production of healthcare knowledge. The workings of power are evident within and between the disciplinary discourses of HTA. The health technology metaphor serves the state by enabling the participation of the agencies of knowledge production required by the vision of a knowledge-based health service. The elasticity of HTA's metaphors enables uneasy partnerships to exist between disciplines and methodologies, such as qualitative and experimental methods, and clinical science and sociology.

In health technology assessments, such as clinical trials, the application of technologies in healthcare provision and the testing of them are conflated for a proportion of health care users. Patients are entered into trials of alternative treatments, and models are built to project the effect of health technologies on real patient populations. Thus HTA defines the NHS, its patients and health professionals, as a research laboratory for the field-testing of the effectiveness and cost-effectiveness of healthcare technologies. The inclusion of 'patient preferences' in health technology experiments (*e.g.* Torgerson *et al.* 1996) suggests that, as consumers, our intimate feelings about treatment or other healthcare choices can be incorporated into health technology. If the exercise of choice itself might affect health, then personal healthcare choices become part of the material from which some health technologies are manufactured.

The different combinations of the variable relations between experiments and laboratories in health technology assessment, such as the multi-centre randomised controlled trial combined with cost-effectiveness analysis, pose methodological and analytic challenges for the sociology of healthcare knowledge. For example, HTA projects construct *ad hoc* laboratories in the healthcare environment, which are not marked by separate physical space, and resources for experimental work are widely dispersed between organisations and disciplines. Analytically, a challenge is presented by the perception that, in HTA, health professionals and patients themselves constitute part of health technologies and the measurement instrumentation employed in experiments.

Health technology assessment provides a locus in which relationships between rational science, scientific uncertainty, trust and authority may be examined. Its processes constitute and define areas of expertise characterised by scientific uncertainty. The negotiable 'boundaries' between acceptable and non-acceptable risk, conventionally if implicitly conceived of as lines of demarcation, might better be re-conceptualised as broad bands where questions of comparative risk to health and budgets are held, pending expert, political and public actions related to scientific evidence. Laboratories harbour uncertainty while engaged in projects to reduce it. National health technology assessment policy exemplifies a rational-planning approach to the dynamics of science and health care. From the perspective of the sociology of scientific knowledge, it is an activity which is open to analysis as a scientific practice, with associated contestable knowledge claims. Contestability of expertise is frequently taken to be one of the defining features of health-related issues in contemporary societies, and it is associated with a decline in trust of medical authorities (Gabe and Bury 1996). However, it is not obvious that this is a clear, unopposed trend across healthcare. The HTA movement, in its engagement of expert disciplines, its research alliances, its encompassing metaphors, its containment of uncertainty, its consultation of public voices, its bias-elimination methodologies, and its use of a rational aggregated voice speaking for the good of the public health, might be seen as part of a dialectical process involving decline in traditional medical authority and a reconstruction of a new framework of scientific authority. We are invited to place trust again in this new form of scientific expertise. It is paradoxical that this invitation to place our trust in science is associated with the radical questioning of more established forms of healthcare knowledge which the health technology assessment movement promotes.

Acknowledgements

I should like to thank the editor and the anonymous referees for comments on previous versions of this chapter, and Caroline Brown for comments and advice.

Notes

1 Since 1991, the NHS has been organised by a division between health authorities – 'purchasers' or 'commissioners' – which contract for health services with 'providers' (called NHS Trusts) in the form of hospitals and community health service agencies.
2 The use of the example of gastric freezing to support the case for assessment is something of a selective reading of the assessment history of this particular operation. It was invented and unevenly diffused primarily in the USA, after testing with apparent success on animals. At least thirty-six trials of the procedure were published in American medical scientific journals during the 1960s, the early studies being at least 'qualified favourable'. With hindsight, many of these studies are regarded as having methodological flaws. The negative results of the only multicentre randomised controlled trial were not published until the procedure had fallen out of favour (Fineberg 1979). Gastric freezing spread in spite of assessments, rather than because of lack of assessment.
3 For example, a multi-interest conference 'Planning national research priorities: foresight and the science base in wealth and health creation', held in Cambridge in 1994, organised and supported by SmithKline Beecham Pharmaceuticals and attended by the NHS Director of Research and Development.

References

Advisory Group on the Ethics of Xenotransplantation (1996) *Animal Tissue into Humans*. London: Department of Health.
Ashmore, A., Mulkay, M. and Pinch, T. (1989) *Health and Efficiency: a Sociology of Health Economics*. Milton Keynes: Open University Press.
Austoker, J. (1989) Walter Morley Fletcher and the origins of a basic biomedical research policy. In Austoker, J. and Bryder, L. (eds) *Historical Perspectives on the Role of the MRC*. Oxford: Oxford University Press.
Bachmann, M. and Nelson, S. (1996) *Screening for Diabetic Retinopathy: a Quantitative Overview of the Evidence, Applied to the Populations of Health Authorities and Boards*. Bristol: Health Care Evaluation Unit, Department of Social Medicine, University of Bristol.
Banta, H.D. and Gelijns, A. (1987) Health care costs: technology and policy. In Schramm, C.J. (ed) *Health Care and its Costs*. New York: W.W. Norton and Co.
Banta, H.D. and Luce, B.R. (1993) *Health Care Technology and its Assessment: an International Perspective*. New York: Oxford University Press.
Bartley, M. (1990) Do we need a strong programme in medical sociology?, *Sociology of Health and Illness*, 12, 371–89.
Beer, G. and Martins, H. (1990) Introduction, *History of the Human Sciences*, 3, 163–75.
Berg, M. (1995) Turning a practice into a science: reconceptualising postwar medical practice, *Social Studies of Science*, 25, 437–76.
Black, N. (1996) Why we need observational studies to evaluate the effectiveness of health care, *British Medical Journal*, 312, 1215–18.

Booth, C.C. (1989) Clinical research. In Austoker, J. and Bryder, L. (eds) *Historical Perspectives on the Role of the MRC*. Oxford: Oxford University Press.

Carter, S. (1995) Boundaries of danger and uncertainty: an analysis of the technological culture of risk assessment. In Gabe, J. (ed) *Medicine, Health and Risk. Sociology of Health and Illness Monograph*. Oxford: Blackwell.

Casper, M.J. and Berg, M. (1995) Constructivist perspectives on medical work: medical practices and science and technology studies, *Science, Technology and Human Values*, 20, 395–407.

Challah, S. and Mays, N.B. (1986) The randomised controlled trial in the evaluation of new technology: a case study, *British Medical Journal*, 292, 877–9.

Cochrane, A.L. (1971) *Effectiveness and Efficiency: Random Reflections on Health Services*. London: Nuffield Provincial Hospitals Trust.

Collins, H. (1984) Researching spoonbending: concepts and practice of participatory fieldwork. In Bell, C. and Roberts, H. (eds) *Social Researching: Politics, Problems, Practice*. London: Routledge and Kegan Paul.

Committee on Technology and Health Care, Assembly of Engineering, National Research Council and Institute of Medicine (1979) *Medical Technology and the Health Care System: a Study of the Diffusion of Equipment-embodied Technology*. Washington, DC: National Academy of Sciences.

Dawson, J., Fitzpatrick, R., Carr, A. and Murray, D. (1996) Questionnaire on the perceptions of patients about total hip replacement, *Journal of Bone and Joint Surgery*, 78-B, 185–90.

Department of Health (1993) *Research for Health*. Leeds: Department of Health.

Department of Health (1995) *Report of the NHS Health Technology Assessment Programme 1995*. London: Department of Health.

Dickersin, K. and Berlin, J.A. (1992) Meta-analysis: state-of-the-science, *Epidemiologic Reviews*, 14, 154–76.

Drummond, M. (1994) *Economic Analysis alongside Controlled Trials*. London: Department of Health.

Elston, M.A. (1991) The politics of professional power. In Gabe, J., Calnan, M. and Bury, M. (eds) *The Sociology of the Health Service*. London: Routledge.

Epstein, S. (1995) The construction of lay expertise: AIDS activism and the forging of credibility in the reform of clinical trials, *Science, Technology and Human Values*, 20, 408–37.

Eysenck, H. (1994) Meta-analysis and its problems, *British Medical Journal*, 309, 789–92.

Fineberg, H.V. (1979) Gastric freezing – a study of diffusion of a medical innovation. In Committee on Technology and Health Care *Medical Technology and the Health Care System: a Study of the Diffusion of Equipment-embodied Technology*. Washington, DC: National Academy of Sciences.

Frankel, S. (1989) The natural history of waiting lists – some wider explanations for an unnecessary problem, *Health Trends*, 21, 56–8.

Gabe, J. and Bury, M. (1996) Halcion nights: a sociological account of a medical controversy, *Sociology*, 30, 447–69.

Gieryn, T.F. (1983) Boundary-work and the demarcation of science from non-science: strains and interests in professional ideologies of scientists, *American Sociological Review*, 48, 781–95.

Good, J.M.M. (1993) Quests for interdisciplinarity: the rhetorical constitution of social psychology. In Roberts, R.H. and Good, J.M.M. (eds) *The Recovery of Rhetoric: Persuasive Discourse and Disciplinarity in the Human Sciences*. London: Bristol Classical Press.

Good, J.M.M. and Roberts, R.H. (1993) Introduction: persuasive discourse in and between disciplines in the human sciences. In Roberts, R.H. and Good, J.M.M. (eds) *The Recovery of Rhetoric: Persuasive Discourse and Disciplinarity in the Human Sciences*. London: Bristol Classical Press.

Hoare, J. (1992) *Tidal Wave: New Technology, Medicine and the NHS*. London: King's Fund Centre.

Hoggett, P. (1991) A new management in the public sector, *Policy and Politics*, 19, 243–56.

Hughes, D. and McGuire, A. (1992) Legislating for health: the changing nature of regulation in the NHS. In Dingwall, R. and Fenn, P. (eds) *Quality and Regulation in Health Care: International Experiences*. London, New York: Routledge.

Irving, M. (1993) *Hunter's Baton*. Oration delivered on the Bicentenary of the death of John Hunter. London: Royal College of Surgeons of England. 14 September 1993.

Jamous, H. and Peloille, B. (1977) Professions or self-perpetuating systems? Changes in the French university hospital system. In Jackson, J. (ed) *Professions and Professionalisation*. Cambridge: Cambridge University Press.

Jennett, B. (1986) *High Technology Medicine: Benefit or Burden?* Oxford: Oxford University Press.

Jones, R. (1995) Why do qualitative research?, *British Medical Journal*, 311, 2.

Knorr Cetina, K. (1992) The couch, the cathedral, and the laboratory: on the relationship between experiment and laboratory in science. In Pickering, A. (ed) *Science as Practice and Culture*. Chicago: University of Chicago Press.

Latour, B. (1987) *Science in Action*. Milton Keynes: Open University Press.

Latour, B. and Woolgar, S. (1979) *Laboratory Life*, Princeton, NJ: Princeton University Press.

Leigh Star, S. (1991) Power, technology and the phenomenology of conventions: on being allergic to onions. In Law, J. (ed) *A Sociology of Monsters: Essays on Power, Technology and Domination*. London: Routledge.

MacKenzie, D. (1989) From Kwajalein to Armageddon? Testing and the social construction of missile accuracy. In Gooding, D., Pinch, T. and Schaffer, S. (eds) *The Uses of Experiment: Studies in the Natural Sciences*. Cambridge: Cambridge University Press.

Macleod, R. and Rehbock, P.F. (1994) *Darwin's Laboratory: Evolutionary Theory and the Natural History of the Pacific*. Honolulu: University of Hawaii Press.

Mays, N. and Pope, C. (1995) Rigour and qualitative research, *British Medical Journal*, 311, 109–12.

Maynard, A. (1997) Evidence-based medicine: an incomplete method for informing treatment choices, *Lancet*, 349, 126–8.

Maynard, A. and Chalmers, I. (1997) *Non-random Reflections on Health Services Research*. London: BMJ Publishing Group.

McDowell, I. and Jenkinson, C. (1996) Development standards for health measures, *Journal of Health Services Research and Policy*, 1, 238–46.

Medical Research Council (1991) *Annual Report 1990–91*. London: Medical Research Council.

Medical Research Council (1992) *Annual Report 1991–92*. London: Medical Research Council.

Medical Research Council (1995a) *Research Developments Relevant to NHS Practice, Public Health and Health Departments Policy*. London: Medical Research Council.

Medical Research Council (1995b) *Scientific Strategy*. London: Medical Research Council.

NHS Centre for Reviews and Dissemination (1996) *Undertaking Systematic Reviews of Research on Effectiveness: CRD Guidelines for Those Carrying Out or Commissioning Reviews*. CRD Report 4. York: NHS Centre for Reviews and Dissemination, University of York.

NHS Executive (1996) *Report of the NHS Health Technology Assessment Programme 1996*. Leeds: NHS Executive.

O'Brien, D.M. and Marchand, D.A. (1982) Politics, technology, and technology assessment. In O'Brien, D.M. and Marchand, D.A. (eds) *The Politics of Technology Assessment: Institutions, Processes, and Policy Disputes*. Massachusetts, Toronto: D.C. Heath.

Pickstone, J.V. (1993) Ways of knowing – towards a historical sociology of science, technology and medicine, *British Journal for the History of Science*, 26, 433–58.

Pinch, T. (1993) 'Testing – one, two three . . . testing!': Toward a sociology of testing, *Science, Technology and Human Values*, 18, 25–41.

Pinch, T., Ashmore, M. and Mulkay, M. (1992) Technology, testing, text: clinical budgeting in the UK National Health Service. In Bijker, W.E. and Law, J. (eds) *Shaping Technology/Building Society*. Cambridge, MA, MIT Press.

Porter, T.M. (1995) *Trust in Numbers: the Pursuit of Objectivity in Science and Public Life*. Princeton: Princeton University Press.

Research and Development Directorate, NHSE (1995) *Management and Support for the NHS Health Technology Assessment (HTA) Programme*. Unpublished tender specification document.

Ryan, M. and Shackley, P. (1995) Assessing the benefits of health care: how far should we go?, *Quality in Health Care*, 4, 207–13.

Sackett, D.L., Rosenberg, W.M.C., Muir Gray, J.A., Haynes, R.B. and Richardson, W.S. (1996) Evidence based medicine: what it is and what it isn't, *British Medical Journal*, 312, 71–2.

Sheldon, T. and Faulkner, A. (1996) Vetting new technologies, *British Medical Journal*, 313, 508.

Smith, R. (1994) Towards a knowledge based health service, *British Medical Journal*, 309, 217-18.

Swales, J.D. (1997) Science in a health service, *Lancet*, 349, 1319–21.

The Cochrane Collaboration (1994) *Handbook Section VI: Preparing and Maintaining Systematic Reviews*. Oxford: Cochrane Collaboration.

Torgerson, D.J., Klaber-Moffet, J. and Russell, I.T. (1996) Patient preferences in randomised trials: threat or opportunity? *Journal of Health Services Research and Policy*, 1, 194–7.

Weatherall, D. (1995) *Science and the Quiet Art: Medical Research and Patient Care*. Oxford: Oxford University Press.

Webster, A. (1994) University-corporate ties and the construction of research agendas, *Sociology*, 28, 123–42.

Notes on contributors

John Abraham is a Lecturer in Sociology at the University of Sussex where he teaches courses in 'Medicine, State and Society' and 'Science, Technology and Society'. For the past eleven years he has researched the testing of medicines in the UK, US and the EU and has published widely on this.

Paul Atkinson is Professor of Sociology in the University of Wales, Cardiff. His research interests include the microsociology of knowledge, the rhetoric of sociological inquiry and ethnographic research methods. Among his recent publications are two books on computing and qualitative data analysis as well as studies of the 'new' genetics and haematology.

Mel Bartley is a contract researcher in medical sociology. She has a special interest in the relationship of physical and mental health to employment, unemployment and living standards for men and women, and in the ways in which our knowledge in these areas is socially constructed.

Claire Batchelor is currently a postgraduate student in the School of Social and Administrative Studies at the University of Wales, Cardiff. She has a Master's degree in Population Studies and has worked as a research associate on various research projects, including the ESRC-funded grant on the discovery of the Myotonic Dystrophy gene.

David Blane is Senior Lecturer in Medical Sociology at Charing Cross and Westminster Medical School, University of London. He teaches general medical sociology to medical students and has carried out research on health inequalities, with a special interest in measurement issues.

Helen Busby studied medical (social) anthropology at Brunel University and is now a Research Fellow with the National Primary Care Research and Development Centre at the Public Health Research and Resources Centre, University of Salford. She is currently working on a project about musculoskeletal disorders in the community, concerned with exploring lay knowledge and choices about using health care.

Alberto Cambrosio is an Associate Professor in the Department of Social Studies of Medicine of McGill University. His work lies at the interface of the sociology of science and medical sociology, with a particular focus on the material cultures of biomedicine (i.e. technological and transplantation) and on the role of imagery in immunological research.

George Davey Smith is Professor of Clinical Epidemiology in the Department of Social Medicine, University of Bristol. His research interests include inequalities in health, the childhood origins of adult disease and sexually transmitted disease prevention.

Mary Ann Elston is Senior Lecturer in Sociology in the Department of Social Policy and Social Science, Royal Holloway, University of London and Course Director for the MSc in Medical Sociology programme there. Her research interests are in the sociology and history of health care occupations, the organisation of biomedical research and popular health movements.

Alex Faulkner is Research Associate in the Department of Social Medicine, University of Bristol. He is a sociologist and health services researcher with interests in health technology, surgical innovation policy, artificial hip technology, prostatic cancer and outpatient health services.

Yoshio Nukaga is a doctoral student in the Department of Sociology, McGill University. He is currently working on a comparative history and ethnography of the use of medical pedigrees in the case of Huntington's Disease in North America and Japan.

Mary-Rose Mueller is a National Institute on Aging Posdoctoral Fellow at University of California, San Francisco. Her areas of research interest are sociology of medical science and work, professional development and social studies of ageing.

Evelyn Parsons is a Research Fellow in the Institute of Medical Genetics and Senior Lecturer in the School of Nursing at the University of Wales College of Medicine. She is currently responsible for the social evaluation of the all-Wales programme of newborn screening for Duchenne Muscular Dystrophy. Her research interests include the social construction of genetic risk and the implications of presymptomatic screening for familial breast cancer.

Anne Rogers is currently Reader in Sociology at the National Primary Care Research and Development Centre, University of Manchester. Her current research interests are in the sociology of mental health and illness, help-seeking and health services utilisation and access to primary care.

Gareth Williams is Professor of Sociology and Deputy Director of the Public Health Research and Resource Centre, University of Salford. He has published widely in sociology and health journals and written or edited several books on chronic illness and disability, lay knowledge of health and illness, and the NHS reforms.

Index

Abbott, A. 73
Abraham, J. 10
actor-network theory 9, 13
actuaries, and mortality measurement 131–2, 135
adverse drug reactions 158, 159–60
age standardisation 132–3, 135–6, 137, 138, 145
AIDS, clinical research into 59, 60, 61–3, 64–9
AIDS Clinical Trials Group (ACTG) 61–2, 64, 73
alliances
  in epidemiology 133–5, 142
  in public health policy 146–7
  in research 184, 192–3
animal protection, in drug testing 173–4
Anspach, R. R. 75
Ashmore, A. 187, 197
Association of the British Pharmaceutical Industry (ABPI) 163, 164
Atkinson, P. 2, 29

Balls, M. 173
Banta, D. 188, 200
Barber, B. 7
Barnes, B. 104
Bartley, M. 153, 186
Bennett, R. L. 37, 40
Berg, M. 3, 4, 10
Billings, P. R. 45
biomedical model 1, 4
boundary work 19, 187
Brownlee, J. 138–9, 146
Burley, D. M. 159
Bury, M. 18, 174

Campbell, J. M. 141
Canada, genetic counselling in 35, 37, 38, 41–2, 51n
case-control study 142, 144, 147
case records 40, 41–2
Casper, M. J. 4, 10
Centre for Medicines Research (CMR) 173
centres of calculation 129, 130, 131, 133–4, 148
Chadwick, Edwin 130–3
chance, in scientific discovery 113, 115, 116, 119, 122
Chubin, D. 154
circulation of medical pedigrees 34, 46–9
civil registration 135
clinical research 57–76
clinical trials 61–3, 158, 159, 202
  science/care dilemmas in 63–9, 73–4
clinical trials certificates (CTCs) 164
clinical trials exemptions (CTXs) 165
closeness, in scientific discovery 115–16
Cochrane, Archibald 189, 194–5
Codman, Ernest 189
Cohen Committee 161–2
collaboration in research 106, 111–12
Collins, H. M. 107, 185
combination of medical pedigrees 34, 42–6
Committee for Proprietary Medicinal Products (CPMP) 170, 171, 172
Committee on the Safety of Drugs (CSD) 161–3
Committee on the Safety of Medicines (CSM) 164
conflict in clinical science *see* science/care dilemma
construction of medical knowledge 8–11, 121
contingent repertoires of accounting 109, 110–11, 119
Cook-Deegan, R. M. 43
co-ordination of institutions, in health technology assessment 192–3
Cornfield, Jerome 141, 142–3, 146
corporatism 155, 166, 171, 172, 175
costs and benefits 200

Darwinists 136
Davis, P. 1
Dawson, J. 197
degenerative disease 138
disciplinary knowledge 187
division of labour 15–16, 121
  in clinical research 57–8
DNA technology 101, 106
Doll, R. 142
drug testing *see* clinical trials; medicines regulation
Dunlop, D. 163

Elston, M. A. 201
empiricist repertoires of accounting 109–10, 116, 119
epidemic disease 5
epidemiology 10, 196
Ethics in Government Act 1978 (USA) 168
European Medicines Evaluation Agency 171
European Union, and medicines regulation 155, 169–72
experimentalism 194–8

experiments 186, 203
expertise 16, 80–1, 83
  professional 90–4
Eyler, J. M. 137
Eysenck, Hans 141, 143, 144

family, and medical pedigrees 15, 34–8
family trees 35, 37–8, 41, 42
Farr, William 134–5, 136–7
Fentem, J. 173
Fisher, R. A. 141, 143–4
Fletcher, J. C. 29
Food and Drug Administration (FDA) 166–9, 173, 175
Foucault, Michel 30
founding fathers 7, 189
Fox, R. 7, 57, 58
Fragile X gene 47–8, 120
FRAME (Fund for the Replacement of Animals in Medical Experiments) 173
Frankel, M. S. 47
Fujiki, N. 37, 44
Fujimura, J. 8, 121

Gabe, J. 18, 174
Gelehrter, T. D. 41
gender
  health care professions and 72–3
  medical science and technology and 12–14
gene research 101–2
  scientists' accounts of 109–20
General Register Office 133, 135, 138
  as centre of calculation 131, 133
generalisability 194–5, 197
genetic counselling 15, 29, 30–51
genetic mutation 120
genetics 5–6, 50, 102–3
Gilbert, N. G. 103, 109
globalisation of medicines regulation 172–4
Glynne, A. 159
GPs 91–5
grants for clinical trials 63, 67, 74
Greenwood, J. 155

health care
  during clinical trials 67–9, 73–4
  quality of 18
  scientific knowledge and 186
  see also science/care dilemma
health economics 197–8
health service professions 7–8
  gender and 72–3
  intraprofessional conflict 74–5
health technology 190–1
health technology assessment (HTA) 19, 183, 184–5, 187, 198–202
  justifying need for 188–92
  research alliances in 192–3
Hill, A. B. 141, 142, 146
Hunter, John 189
Huntington's Disease 42, 44, 111

information 40, 43
  about medicines 166, 167, 171–2, 175, 176
Institute for the Struggle against the Dangers of Tobacco 140
insurance industry, and mortality risk assessment 134–5
interest representation 155
International Conference on Harmonisation (ICH) 172, 174

Japan, genetic counselling in 35, 37, 38, 44–5, 51n.
Jasanoff, S. 155

Kanazawa, I. 44
Kennedy, Donald 168
Knorr Cetina, K. 186
knowledge of joint pains
  and experience 80–1, 82
  lay 84–5, 90, 93–4
  specialist 83–4
  see also medical knowledge; scientific knowledge

laboratories 3, 186, 203
  and production of medical knowledge 121
laboratory pedigrees 45–6
Lantos, J. 64
large pedigrees 42–4
Latour, B. 8, 129, 133–4
Laurence-Moon-Biedl syndrome 44
lay knowledge 80–1, 83, 84–5, 90, 93–4
  devaluation of 82–3
life expectancy 136, 137
  measurement of 130–3, 136–7
life tables 131, 135, 137
Litchfield, J. T. 159
lung cancer, cause of 140–5

MacKenzie, D. 186
maps, family trees as 37–8
measurement of mortality risk 127, 129, 133, 139–47
medical knowledge 83–4
  construction of 8–11, 121, 123
  and health technology assessments 192
  lay responses to 90–3
  politics and 154
medical genetics 102–3
Medical Officers of Health 137
medical pedigrees 29–32, 49–51
  genetic counselling and 32–3
  production of 33–49
medical records 66

Medical Research Council 138, 140, 145, 146, 193
medical scientists, and drugs regulation 165, 167, 169
medicines 1
lay/expert understanding of 11–12
regulation of *see* medicines regulation
Medicines Act 1968 163
Medicines Control Agency (MCA) 165–6
medicines regulation 174–6
Europeanisation of 169–72
globalisation of 172–4
in UK 161–6
in USA 166–9
Merton, Robert 7
molecular genetics 5–6, 50
Mulkay, M. 103, 109
multi-disciplinarity 189–90, 192, 194–8
musculoskeletal disorders, lay and biomedical understandings of 79, 83–95
myotonic dystrophy 17, 43
identification of gene 101, 104, 106–7
research into 105–20, 121

narrative accounts 103, 109–20
National Institutes of Health 141, 146
Neison, F. G. P. 131–2, 145
new reproductive technologies 14
NHS 19, 183
research knowledge 184–5, 192–3
nurses in clinical trials 57–8, 62–3, 67–9
management of science/care dilemma 70–2
and patient-volunteers 64, 65–6, 74–5
sex role socialisation 72–3

Oakley, A. 58, 69
occupational inequalities in health 139, 145
odds ratio 18, 142–5, 145–6
Ogle, William 136, 138
Olson, D. R. 38
organisation of medical practice 75
osteoarthritis 84, 90, 92
outcome measures 196–7, 198, 199

participation in clinical trials 65–6
patient-volunteers 67
medical care for 67–9
status of 63–7
pharmaceutical industry, and medicine regulation 158–9, 161, 164, 165, 166, 167–8, 171, 173
physicians in clinical research 57–8, 62, 67–9, 73, 74, 75
management of science/care dilemma 70–2
and patient-volunteers 63–4, 65, 66
Pickstone, J. V. 194
Pinch, T. 187, 191
power of GPs 91
predictability in scientific discovery 108, 111 114–15, 116, 122
prediction 104, 122–3
primary transcription of medical pedigree 33, 34–8
product licensing 162–3
professional expertise 94
lay responses to 90–3
projection 194–8
protocols in clinical trials 63, 65–6, 67–9, 71
public deficit model 11
public health debate 131–40
public understanding of science 11, 82
publication of medical pedigrees 46–9

qualitative research 196, 198
quality of life measures 197, 198

Raffell, S. 40
randomised controlled trials 194–5
realism 10, 154
regulatory science 18, 154–5
reproduction 13–14
Restivo, S. 154
rhetoric, and health technology assessment 187, 188, 189, 191–2, 198
rheumatology 84
Richards, M. R. 4
risk of disease 141–5, 199
assessment of 18, 199–200
Ronit, K. 155

Salsburg, D. 159
Schroeder, R. 11
science
and health care 3, 4, 6, 57–76, 189
lay/expert understanding of 11–12
women in 13
science/care dilemma 59, 63–9
management of 70–2
social organisation and 72–6
scientific discovery 102–4
accounts of 109–20
scientific knowledge 81–3, 186
see also sociology of scientific knowledge
Seabrook, Jeremy 83–4
Searle, John 10
secondary transcriptions of medical pedigrees 33–4, 39–42
sex role socialisation 72–3
Simpson, S. D. 47
smoking, and lung cancer 141–3
social constructionism 8–10, 187
social organisation, and work-related conflict 72–6
sociology of health and illness 81–3
sociology of medicine 3–4, 6–14, 186–7
and sociology of science 153–4

sociology of science 6–14, 153–4
sociology of scientific knowledge (SSK) 8–11, 81–3, 154, 186–7
  strong and weak programmes of 8, 153
Standardised Mortality Ratio (SMR) 17–18, 130, 136, 138, 145, 147–8
  criticisms of 138–9
Standing Group on Health Technology (SGHT) 184–5, 193
Star, S. L. 16
statistics 133, 134
Steinhaus, K. A. 37
Stocking, Barbara 189
Stocks, P. 141
Stolley, P. D. 143
Strauss, Anselm 6
Swales, J. D. 189
systematic review 195
Szreter, S. 129, 138

technological testing 186–7
technology 9, 188, 190
  social constructions of 9–10
  see also health technology
Teich, A. H. 47
textbook pedigrees 49
toxicological tests 158, 159
translation model of science 128–30
uncertainty
  drug testing and 158, 159–60, 173
  in scientific discovery 111, 113, 114, 116–17
United Kingdom, medicines regulation in 161–6
United States of America
  clinical trials in 61–3
  medicines regulation in 166–9
University College Hospital 2
University College London, biomedical research centre 1, 2

validation 197
Vandenbrouke, J, P. 143–4
village pedigree 44–5
'voices of medicine' 29

wear and tear 83, 86, 88–90, 91
Weatherall, D. 183
Weitz, R. 74
Williams, G. 12
Williams, S. J. 10
Wilson, G. M. 161–2
Wolff, S. 144, 145
work, musculoskeletal disorders and 88–90
Wynne, B. 12

Yule, G. U. 139, 145

Zussman, R. 75